Other Worldwatch publications available from Earthscan

Vital Signs: The Trends that are Shaping our Future
1996–7 edition £12.95 paperback ISBN 1 85383 367 3

State of the World
1997 edition £12.95 paperback ISBN 1 85383 427 0

Saving the Planet: How to Shape an Environmentally Sustainable Economy
Lester R Brown et al
£8.95 paperback ISBN 1 85383 133 6

How Much is Enough? The Consumer Society and the Future of the Earth
Alan Thein Durning
£8.95 paperback ISBN 1 85383 134 4

The Last Oasis: Facing Water Scarcity
Sandra Postel
£9.95 paperback ISBN 1 85383 148 4

Full House: Reassessing the Earth's Population Carrying Capacity
Lester R Brown and Hal Kane
£9.95 paperback ISBN 1 85383 251 0

Power Surge: A Guide to the Coming Energy Revolution
Christopher Flavin and Nicholas Lenssen
£10.95 paperback ISBN 1 85383 205 7

Who Will Feed China? Wake-Up Call for a Small Planet
Lester R Brown
£9.95 paperback ISBN 1 85383 316 9

Tough Choices: Facing the Challenge of Food Scarcity
Lester R Brown
£9.95 paperback ISBN 1 85383 432 7

Fighting for Survival

FIGHTING FOR SURVIVAL

Environmental Decline, Social Conflict, and the New Age of Insecurity

Michael Renner

The Worldwatch Environmental Alert Series
Linda Starke, Series Editor

EARTHSCAN

Earthscan Publications Ltd, London

Worldwatch Database Disk

The data from all graphs and tables contained in this book, as well as from those in all other Worldwatch publications of the last two years, are availaable on disk for use with IBM-compatible or Macintosh computers. This includes data from the *State of the World* and *Vital Signs* series of books, Worldwatch Papers, *WorldWatch* magazine, and the Environmental Alert series of books. The data are formatted for use with spreadsheet software compatible with Lotus 1-2-3 version 2, including all Lotus spreadsheets, Quattro Pro, Excel, SuperCalc, and many others. for IBM-compatibles, a 3½-inch (high density) disk is provided. To order, please contact Earthscan at the address below with a cheque or crediit card authorisation for £70 plus 10% p+p.

First published in the UK in 1997 by
Earthscan Publications Ltd

First published in the US in 1997 by
WW Norton and Company Inc
500 Fifth Avenue, New York, NY 10110

A catalogue record for this book is available from the British Library

ISBN: 1 85383 433 5 (paperback)

Printed and bound by Biddles Ltd, Guildford and King's Lynn
Cover design by Declan Buckley

For a full list of publications, please contact
Earthscan Publications Ltd
120 Pentonville Road
London N1 9JN
Tel: 0171 278 0433
Fax: 0171 278 1142
email: earthinfo@earthscan.co.uk
http://www.earthscan.co.uk

Earthscan is an editorially independent subsidiary of Kogan Page Ltd and publishes in association with WWF-UK and the International Institute for Environment and Development.

Contents

Acknowledgments *7*

Foreword *11*

1. The Transformation of Security *17*

I. Sources of Stress *33*

2. Environmental Stress *35*

3. Conflict Over the Environment *52*

4. Inequality and Insecurity *76*

5. People on the Move *97*

6. Vicious Circles: Two Case Studies *114*

II. Security in the Twenty-First Century *133*

7. A Human Security Policy *135*

8. Enhancing International Peace Capacity *154*

9. A Human Security Budget *173*

10. A Global Partnership for Human Security? *189*

Notes *199*

Index *231*

Acknowledgments

When the time came to select an appropriate cover illustration for this book—about halfway into the writing process—the task provided a welcome change of pace from months of sifting through books and articles, and of sitting in front of a computer screen. Yet it seemed an almost impossible assignment: how to condense the complex social, economic, and environmental pressures that are at the core of the analysis in *Fighting for Survival* into a single visual image?

Luckily, my colleagues and I came across a picture that seemed just right. Everyone in the Worldwatch office was struck by the simple yet compelling image of a young boy confronting the massive deforestation in his native Madagascar. It shows the harsh realities in this island nation, and yet the bright colors provide some inexplicable hope. In this sense, the illustration aptly reflects the message of this book: although humanity is confronting grave threats, there are solutions. What made this illustration so attractive is that it puts a human face on the issues covered here.

This book was researched and written in the relatively short space of about one year. But it is the culmination of a much longer process that reaches back to the early eighties. I have drawn on the various interests that I have pursued over time, both during my nine years at Worldwatch and some additional years elsewhere, and sought to weave together topics that are all too often regarded as separate and unrelated.

I would like to express my sincere gratitude to all my colleagues at Worldwatch Institute for the many ways in which they have supported this project. A series of informal discussions early on were critical in framing the issues, tailoring the core message and giving me the necessary boost to push ahead. Just as important were the countless bits and pieces of information or insights that my colleagues kept passing to me along the way. I want to extend particular thanks to Gary Gardner, who went out of his way to help me think through some of the more vexing issues, and to Aaron Sachs, whose work on the link between environment and human rights has greatly influenced my thinking. I also greatly appreciate the help of Lori Baldwin, the Worldwatch librarian, who not only tirelessly tracked down articles and reports I requested but also pointed out materials I was unaware of.

Particular thanks are due to Christopher Flavin for providing steady guidance and advice throughout this endeavor. This book would not be what it is without Chris's ability to review drafts on extremely short notice and provide stimulating feedback. I am similarly indebted to Hilary French for her willingness to review several draft chapters despite a busy schedule. I also benefited from the comments of Ken Epps at the Project Ploughshares in Waterloo, Canada; P.J. Simmons of the Woodrow Wilson Center in Washington, D.C.; and Jeffrey Boutwell of the American Academy of Arts and Sciences in Cambridge, Massachusetts.

Because this book builds on earlier work, I would like to acknowledge a number of individuals who have been pivotal to spurring my interest in the topics covered and from whom I have learned a great deal. Among others, they include Robert Johansen at Notre Dame University in Indiana; Seymour Melman, Professor Emeritus at Columbia University; Greg Bischak of the National Commission for Economic Conversion and Disarmament; Herbert Wulf of the Bonn International Conversion Center; Inge Kaul, until recently Director of the Human Development Report Office at the U.N. Development Programme; Ruth Caplan of the Economics Working Group; and Sylvanus Tiewul, whose untimely death was a great loss not just for me but for the United Nations, where he served as one of the more inspiring international civil servants.

Finally, Worldwatch researchers do not live on facts alone. The different forms of moral support and encouragement that I received along the way, from Worldwatch colleagues, my family, and others, were critical in overcoming doubts and obstacles. I especially want to thank Elena Pérez for her cheerfulness, which helped me keep things in perspective and allowed me to cope with some of the more challenging parts of writing a book.

I dedicate this book to my children, Paul and Judith, and to children all over the world—the generation for whom human security and sustainable development will finally have to become more than mere slogans.

Michael Renner

Worldwatch Institute
1776 Massachusetts Ave., N.W.
Washington, D.C. 20036

June 1996

Foreword

In October 1977, Worldwatch Institute published a paper by Lester Brown entitled *Redefining National Security.* The idea was a radical one at the time—that a nation's security depended more on the health of its economy, its natural resource base, and its people than on its military preparedness. The cold war seemed like it would never end, and battles for the loyalties of people in Vietnam and many other "hot spots" were still on everyone's minds.

Nearly 20 years later, Michael Renner returns to the topic of security. The environmental deterioration, population pressures, and economic insecurity that Lester wrote about in 1977 have worsened, yet concepts of national security remain narrowly focused on military concerns. The cold war is over, and soon schoolchildren will have to turn to history books to find out about the Berlin Wall, the Evil Empire, the Iron Curtain, and puppet states. But the world continues to spend something on the order of $800 billion a year on staying ready to fight a war—the equivalent of the income of nearly half the world's people.

There are indeed many battles still to be won, but they are not on the traditional fields. The new battlegrounds are found where biological diversity is being lost at record rates, where children die needlessly of diarrhea or measles, where soil erosion or deforestation rob the land of its productivity, where people cross borders in search of better jobs, better soils, better grazing lands. These are the sources of national insecurity that Lester wrote about before the cold war ended, and that have only worsened since then.

Many of these pressures and insecurities can be found in Chiapas and Rwanda, which Michael provides as case studies in Chapter 6. The pressures of population growth, environmental decline, and economic insecurity in these two regions can be found in communities and societies the world over. As Michael notes, "military might—the traditional answer to societal challenges—is practically irrelevant" in these situations.

In Part II, Michael considers the policies needed for a new approach to security, one based on human well-being and development. And he looks at how to pay for these essential investments in a world already stretched thin by national and global financial commitments. No less than a fundamental reexamination of the assumptions that have long guided national security policies is required, Michael concludes.

Fighting for Survival is the eighth book in the Environmental Alert Series, which was launched in 1991. (See complete list of the series on page 2.) Each volume deals with a particular aspect of the world's progress toward a sustainable society—or lack of progress, in all too many cases. Your comments and suggestions for future topics are always welcome.

Linda Starke, Series Editor

Fighting for Survival

1

The Transformation of Security

For 50 years, sustained by the cold war, "security" has been defined primarily in military terms. While the East-West ideological and military standoff divided much of the planet into two hostile camps, many issues of the day were subordinated to one overriding goal: striving for global supremacy. Backed by doomsday nuclear arsenals, the cold war adversaries were locked in mortal competition.

But now that the cold war has faded away, a very different struggle for survival is emerging. It is becoming clear that humanity is facing a triple security crisis: societies everywhere have to contend with the effects of environmental decline, the repercussions of social inequities and stress, and the dangers arising out of an unchecked arms proliferation that is a direct legacy of the cold war period.

We are at a historic juncture in our understanding of security. The cold war represented the most extreme expression of "national security"—states' desire to protect their borders and territories from foreign invasions, which led over the centuries to the creation of ever larger standing armies and the development of ever more sophisticated weapons.

The emerging issues of the post–cold war era, by contrast, point to a different meaning of security that is much closer to people's tangible concerns. As the 1994 *Human Development Report* points out: "It will not be possible for the community of nations to achieve any of its major goals—not peace, not environmental protection, not human rights or democratization, not fertility reduction, not social integration—except in the context of sustainable development that leads to human security." Concerns about "human security" are as old as human history, yet they are now magnified by the unprecedented scale of environmental degradation, by the presence of immense poverty in the midst of extraordinary wealth, and by the fact that social, economic, and environmental challenges are no longer limited to particular communities and nations.[1]

The cold war, then, can be seen as a relatively brief interlude, a curious historical diversion that distracted our energies from the most basic threats to human society. Unfortunately, a lasting impact of that period is the unparalleled and largely uncontrolled worldwide availability of arms of all calibers. Under these conditions, the temptation to use firepower instead of negotiating skills to resolve conflicts is strong.

The cold war's rigid bipolarity has fallen by the wayside, making room for a more multipolar world in which countries do not automatically rally behind a leader, in which constellations of power and interest seem more transient, and in which diverging interests or rivalries are resurfacing even among old allies. But it is also a world devoid of the sense of mission and destiny that sustained the cold war at high levels of intensity. In large part, this may be due to the pervasive lack of vision.

As Victoria Holt of the Council for a Livable World Education Fund put it, "even the name for our time is elusive; we remain in the 'post–Cold War era,' a period best defined by what is no longer." The cold war structure has not been replaced by any coherent set of multilateral policies, arrangements, and institutions. And it is difficult to marshal the political support and resources necessary to respond to "nontraditional" challenges.[2]

The world has always been more complex than it seemed through the one-dimensional lens of cold war priorities. Yearning for the predictability they had grown accustomed to over the past half-century, however, many policymakers and pundits perceive the world to be suddenly more disorderly, even chaotic. A metaphor used by James Woolsey, former head of the U.S. Central Intelligence Agency, is telling: likening the demise of the Soviet Union to "slaying the dragon," Woolsey compared the future to living in a jungle inhabited by a bewildering variety of poisonous snakes.[3]

* * * *

Woolsey's snake-filled jungle did not emerge overnight. The world already experienced a transformation of conflict during the cold war that it was too distracted to notice: a shift from war between sovereign states to fighting within societies, so that armed conflict conforms less and less to the preoccupations with fending off foreign invasions that are the concerns of traditional national security doctrines.

Since the end of World War II, there have been at least 130 wars, killing more than 23 million people directly and another 20 million through famine and other war-related disruptions. Whereas the number of major wars—killing at least 1,000 persons—stood at around a dozen in any given year during the fifties, and rose no higher than 20 a year during the sixties and seventies, it surged at the beginning of the eighties to more than 30, where it has remained ever since. (See Figure 1–1.)[4]

But far from the traditional image of war—national armies clashing on a well-defined battlefield—violent conflict today increasingly involves protagonists within rather than between countries. The "battlefield" can be anywhere, and the distinction between combatants and noncombatants is blurred.

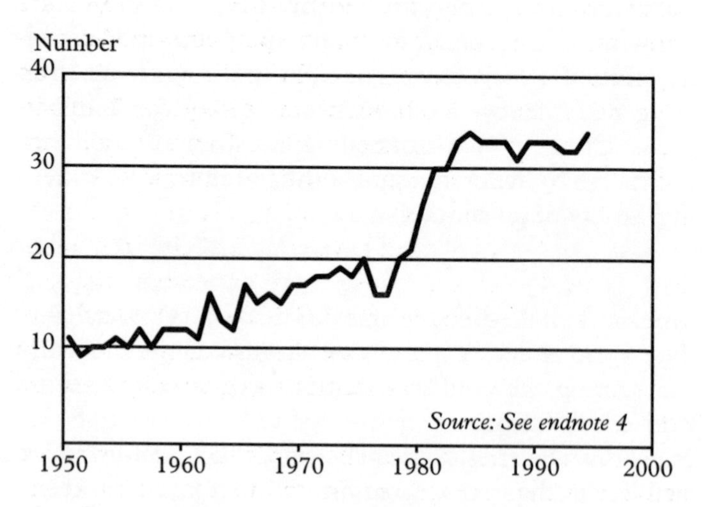

FIGURE 1–1. *Armed Conflicts with More Than 1,000 Deaths, 1950–94*

Since 1945, more than three quarters of warfare has been of an internal nature; in the nineties, virtually none of the major armed conflicts has been unambiguously country-against-country. The fighting is done as often by guerrilla groups, paramilitary forces, and vigilante squads as by regular, uniformed soldiers. Civilians are often not accidental victims but explicit targets of violence, intimidation, and expulsion. Humanitarian relief supplies are frequently delayed and used as bargaining chips in these unconventional wars. Whether through direct violence and terror tactics or conflict-induced starvation and disease, civilians constitute a rapidly growing share of the casualties. By some rough estimates, they accounted for half of all war-related deaths in the fifties, three quarters in the eighties, and almost 90 percent in 1990.[5]

It is a terrible fact that in these conflicts that shred communities and rip apart entire societies, children are transformed into both victims and perpetrators. According to UNICEF, during the eighties 2 million children were killed in war, 4–5 million were disabled, 12 million were left homeless, and some 10 million were psychologically traumatized. In countries as diverse as El Salvador, Sierra Leone, and Sri Lanka, government armies and insurgent forces alike have recruited children, often forcibly. Although the statistics are understandably unreliable, there were an estimated 200,000 child soldiers in the late eighties worldwide. Liberia alone currently has some 20,000 child combatants—a quarter of all those who are fighting.[6]

As many countries may be bordering on war as are actually engaged in it. Dan Smith, director of the International Peace Research Institute in Oslo, points out that in many countries, "political violence [is] widespread and even endemic without quite meriting the name war." From Haiti to the Philippines, highly inequitable social and economic conditions remain in place that trigger cycles of uprisings by the disadvantaged and oppression by the ruling elites. Some countries are facing generalized lawlessness and banditry—whether by marauding ex-soldiers (in several African nations), drug cartels (in Colombia), or various forms of organized crime (in Russia). Analysts are observing a growing privatization of security and violence—in the form of legions of private security guards, the proliferation of small arms among the general population, and the spread of vigilante and "self-defense" groups. A culture of violence—in which violent responses to social problems are the norm—has taken root in many countries.[7]

The post–cold war era is increasingly witnessing a phenomenon of what some have called "failed states"—the implosion of countries like Rwanda, Somalia, Yugoslavia, and others. But they are only the most explicit examples of the pressures and vulnerabilities of the current era—victims of an array of underlying forces that many other countries are subjected to but have managed, for the time being at least, to cope with more successfully.[8]

Several other countries are among the ranks of what Professor James Rosenau of George Washington University calls "adrift nation-states"—countries that "have lost their moorings and may well be moving toward the edge of failure." Among these are, for different reasons, Nigeria, Venezuela, Zaire, and several of the former Soviet republics. One or more of the following conditions is likely to be present: the economy is being depleted; the state is unable to provide anything like adequate services to its citizens; grievances are disregarded and political dissent is repressed; the social fabric is unraveling;

and the political system is unable to cope with growing tensions among different ethnic groups, regions, and classes, or it plays different groups off against each other in an effort to prolong its rule.[9]

The outcome in the case of these unsettled nation-states is by no means preordained: it may be prolonged drift, a gradual revitalization of society, establishment of an authoritarian regime that will crack down hard on any sign of opposition in an attempt to "hold the country together," or a splintering of society.

The term "failed state" implies an unfortunate judgment of finality—conveniently allowing the international community to wash its hands of any responsibility. The outbreak of civil wars and the collapse of entire societies is now routinely being ascribed to the resurfacing of "ancient ethnic hatreds" revolving around seemingly irreconcilable religious and cultural differences. Professor Samuel Huntington of Harvard University went so far as to postulate a coming "clash of civilizations"—ethnically motivated communal violence writ large. This description bears the marks of inevitability, and so forecloses any rational analysis of the roots and origins of contemporary conflicts.[10]

Of course, ethnic tensions do play some role. Some 40 percent of all countries have populations from five or more different "nations." Roughly half of the world's countries have experienced some kind of interethnic strife in recent years. A 1993 study for the U.S. Institute of Peace found 233 minority groups at risk from political or economic discrimination. These groups encompassed 915 million people in 1990, about 17 percent of the world's population. Only 27 of the 233 groups "have left no record...of political organization, protest, rebellion or intercommunal conflict since 1945." Yet a multicultural society need not involve conflict. Tanzania, for example, is home to many different peoples—Zigula, Yao, Sukuma, Haya, Chagga, Asians, and Arabs—with little evidence of tension among them.[11]

Where ethnic tensions do exist, however, they did not arise in a vacuum. One of the continuing legacies of colonial and imperial rule is that boundaries are often arbitrary—drawn not to reflect local realities, but to serve the purposes of the imperial masters. As a result, people of the same culture, language, or ethnicity often found themselves separated by international borders and grouped with people of other backgrounds and origins, irrespective of whether they had previously coexisted peacefully, been at odds, or had no significant contact at all. To steady their rule, colonial administrations typically favored one local group, often a minority, over others—Tamils over Sinhalese in Sri Lanka, Tutsi over Hutu in Rwanda and Burundi, Christian Arabs over Sunni and Shia Moslems in Lebanon—generating a fatal resentment.[12]

Following independence, civic life in many of these states continued to be split along ethnic lines, with one group ruling at the direct expense of the other. Given severe economic underdevelopment and undemocratic, often repressive patterns of governance, the competition for power and resources among contending groups became intense. In light of the vulnerable status of minorities in multiethnic states, it is no surprise that separatist sentiments abound. A 1990 study for the U.S. Institute of Peace found that more than half of 179 minority groups in the Third World feel disadvantaged; although some might be content with pressing for domestic changes in their favor, many have separatist leanings.[13]

On the surface, many conflicts do seem to revolve around ethnic, religious, cultural, or linguistic divisions, and these divisions will likely dominate the perceptions of the protagonists. Yet to gain a better understanding of the nature of these conflicts, we need to look beyond the easy excuse of "ancient hatreds" and "tribal bloodletting" to detect the underlying stress factors that help cause the fighting. As Chapters 2, 3, and 4 show, disputes are often sharpened or even triggered by glaring social and economic inequities—explosive conditions that are exacerbated by the growing pressures of population growth, resource depletion, and environmental degradation.

Disparities in wealth and power are growing both within countries and among them, as the rich are gaining at the direct expense of the poor and the middle classes. Many societies are far better at stimulating economic growth than at ensuring a fair distribution of the benefits, leading to growing discontent and polarization among economic classes and communities. In rural areas, unequal land distribution leaves many millions of people struggling to eke out a living, and often forces them to migrate to marginal, inhospitable areas that cannot sustain them for long. In urban areas, as employment is becoming more precarious with increasing use of labor-saving technologies and a continuing corporate search for cheap pools of labor, competition over jobs is intensifying within and among different countries.

Environmental degradation is now of such magnitude that it threatens human sustainability. The planet's regenerative capacity is increasingly being taxed to the limits. Deforestation, soil erosion, desertification, and the overexploitation and pollution of water resources undermine the natural support systems that human life depends on, reduce natural carrying capacity, and increase the competition for nominally renewable yet scarce resources. Throughout many parts of the developing world, rapidly expanding human populations are outstripping the carrying capacity of the local resource base. Environmental degradation and resource depletion are triggering or aggravating internal and international conflict, and are likely to become even more important in future years as climate change exacerbates the situation.

Together these conditions can form a powerful blend of insecurities; they turn rapidly growing numbers of people into migrants or refugees, and the magnitude and speed of these population movements in turn makes them a factor in generating conflict. (See Chapters 5 and 6.) Accompanied by weak, nonrepresentative political systems that are increasingly seen as illegitimate and incapable of attending to people's needs, these pressures can lead to the wholesale fragmentation of societies. As people turn to ethnic, religious, or other group-based organizations for assistance, protection, and identity, relations with other groups often deteriorate.

Bosnia, Rwanda, and Somalia have shown that the abundance and easy availability of arms can turn social and political upheaval into a violent disintegration of entire countries, triggering devastation on a massive scale. Although the public impression of the Rwandan conflict, for example, is mainly one of machete-wielding individuals on a rampage, the killing was in fact also conducted with machine guns, grenades, mortars, and landmines purchased from France, Egypt, South Africa, and another dozen or so arms suppliers that rushed in "like vultures to a carcass," as Stephen Goose and Frank Smyth of the Human Rights Watch Arms Project wrote in *Foreign Affairs*. The now-deposed Hutu Rwandan government admitted that it bankrupted the economy to pay for imported weapons.[14]

★ ★ ★ ★

The social, economic, and environmental trends that are key to human security are increasingly being shaped not only by the fragmentation implied in the rise of "tribalism," but also by globalization. Trade, investment, travel, and communications tie countries and communities more closely together. Although the nation-state is far from being eclipsed, countries and national governments have less and less ability to shape their own destinies. The meaning of borders, community, and sovereignty is in flux, and that in turn makes national (as opposed to global) security a more tenuous concept.

As James Rosenau observes, "In the political realm, ...authority is simultaneously being relocated upward toward supranational entities, sideward toward transnational organizations and social movements, and downwards toward subnational groups and communities. In the economic realm, production and distribution facilities are being reorganized in both larger economic units and localized regions. In the social and cultural realms, loyalties and values are being reoriented toward both global norms and neighborhood concerns."[15]

Globalization is proceeding very unevenly. Its economic aspects are highly beneficial for some countries, some companies, and some communities—those that are best able to take advantage of greater integration. By contrast, the vast majority of countries and their populations are relatively powerless to affect the central workings of the emerging global system and often highly vulnerable to it. This is particularly true for countries that rely on exports of commodities for a large share of their gross national product and hence their economic survival, yet have precious little influence over the markets that determine the prices their products will fetch.

Within these countries, export-oriented businesses—farmers, resource extractors, and others—are frequently being given priority in the allocation of land, water, credit, and other inputs or forms of assistance over the numerically much larger producers serving domestic markets. At the same time, domestic producers are increasingly being squeezed by cheap imports.

Economic globalization is now principally a corporate-driven process, going hand in hand with privatization, deregulation, and the erosion of the social welfare state. Given the relative ease of relocating factories and shifting investment resources across the planet, the pressure on communities and countries to remain competitive and offer an inviting investment climate is tremendous. In part, this means downward pressure on wages, and a trend toward a low-common-denominator world with regard to working conditions, social welfare, and environmental regulations. Increasingly, these pressures affect even the better-off communities and the well-trained workers.

In the absence of strong rules and norms, globalization could turn into a free-for-all, an intensifying competition among communities worldwide over jobs, income, and economic well-being. By generating deep apprehension and feelings of insecurity, the very unevenness and uncertainty of the globalization process is in itself becoming a source of conflict.

Although global integration also holds promise, there is an enormous gap between the rapid extension of boundary-crossing activities and efforts to create effective, democratic structures to deal with the consequences of vastly increased interdependence and to shape the globalization process so that it benefits human populations across the planet more broadly. While national sovereignty is becoming more circumscribed, global governance structures remain weak. Global policymaking has become a necessity before it has become a reality.

★ ★ ★ ★

The phenomena of globalization and fragmentation and the nature of the social, economic, and environmental pressures worldwide call for a fundamentally different understanding of the meaning of security—Who is to be secure, and by what means?—and hence for a new set of priorities.

Conditioned by a worldview that largely equates security with military strength, traditional analysts tend to regard emerging issues simply as new "threats" to be deterred. By subsuming these new issues under the old thinking of national military security, efforts to address them in effect become militarized. Hence, weapons proliferation is countered by developing new weapons for preemptive raids on foreign arms facilities instead of by promoting disarmament; refugees are seen as menacing hordes to be intercepted on the high seas instead of as people forced from their homes by poverty; environmental degradation is seen as simply another item in which national interests are to be protected against those of other nations instead of acknowledging the common challenge; and the proliferation of drugs is tackled through the military eradicating cocaine crops instead of through efforts to provide alternative livelihoods for desperate peasants.

But many sources of conflict are simply not amenable to any military "solution." Poverty, unequal distribution of land, and the degradation of ecosystems are among the most real and pressing issues undermining people's security. Soldiers, tanks, or warplanes are at best irrelevant in this context, and more likely an obstacle. The military absorbs substantial resources that could help reduce the potential for violent conflict if invested in health, housing, education, poverty eradication, and environmental sustainability.

The past two decades have witnessed a series of efforts to reconceptualize the meaning of security. Recognizing that competitive national security policies had yielded international insecurity, in the early eighties the Independent Commission on Disarmament and Security Issues chaired by Swedish Prime Minister Olof Palme embraced the concept of "common security"—the argument that in order for a state to be secure, its opponents must feel secure. In the eighties and nineties, additional reconceptualizations questioned whether state security was the proper focus and argued that environmental and other nonmilitary factors are at least as important as military ones. The *Human Development Report* produced by the U.N. Development Programme has woven together the different strands of these redefinition efforts and coined the term human security, which is used throughout *Fighting for Survival.*[16]

The twentieth century has seen the pursuit of "national security" elevated to near theological levels; modern military technology has dramatically increased the destructive power of weaponry, the range and speed of delivery vehicles, and the sophistication of targeting technologies. Yet arms ostensibly designed to enhance security increasingly imperil humanity's survival. We live in what is the most violent time in human history: the twentieth century accounts for 75 percent of all war deaths inflicted since the rise of the Roman Empire.[17]

As we approach the twenty-first century, the key question is whether we will see a continuation of the current era, or a departure in security policies that focuses less on the symptoms and more on the root causes of conflict and insecurity, less on the dangers from abroad and more on the perils from within.

An understanding of security consonant with the realities of today's world requires a shift from conflict-laden to cooperative approaches, from national to global security. Instead of defense of the status quo, human security calls for change and adaptation; instead of a fine-tuning of arms and recalibration of military strategies, it calls for demilitarization, conversion of war-making institutions, and new priorities for sustainable development. The dimensions of this new understanding of security are discussed in Part I of this book. Current and needed initiatives to improve our efforts to fight for survival are considered in Part II.

I

Sources of Stress

2

Environmental Stress

The defining feature of the harbor city of Aral'sk in Kazakstan is gone: it was once situated at the northernmost tip of the Aral Sea, but now Aral'sk is 30 kilometers from the receding shoreline. The Aral Sea has lost 75 percent of its volume and 56 percent of its surface expanse since 1960. Government planners had designated the area to be the Soviet Union's cotton bowl; the two rivers feeding the Aral Sea, the Amu Dar'ya and the Syr Dar'ya, were tapped to supply water to huge irrigation systems. Widespread salinization and waterlogging—due to grossly inefficient irrigation—led to spectacular crop yield losses. The Aral fish catch has dropped from 40–50,000 tons to zero, wiping out 60,000 jobs in the fishing industry. As the Aral Sea has shrunk, winds pick up dust, salt, and toxic chemicals from the sea's exposed bottom and deposit them over a wide area. These airborne toxics are contaminating crops and endangering public health.[1]

As the situation in Central Asia illustrates, the unplanned and rapid depletion of natural systems is an important source of insecurity and stress in many societies, whether in the form of reduced food-growing potential, the worsening health of residents, or diminished habitability. Although desertification, soil erosion, deforestation, water scarcity, and the decline of fisheries are worldwide phenomena, some regions are more severely affected than others. The stress is most pronounced in regions that encompass fragile ecosystems (such as arid or semiarid zones) and that have an economy heavily geared to agriculture. Unchecked, environmental degradation has the potential to impoverish people and undermine the long-term habitability of an area. In extreme cases, natural support systems may be weakened so severely that people have little choice but to move.

★ ★ ★ ★

A major threat to the economic well-being of many countries is land degradation—principally through the plowing of highly erodible land, the drawing down of water tables through overpumping for irrigation, the salinization of irrigated land, the overgrazing of rangelands, and the loss of arable land, rangeland, and forests to expanding urban and industrial needs. Although the immediate reasons may be found in inappropriate practices or inefficient technologies, land degradation often is the result of social and economic inequities.

Landless and near-landless peasants are being forced onto marginal lands by unequal land distribution, the lack of secure land tenure, the marginalization of small-scale agriculture by cash-crop operations, the conversion of land to cattle ranching, and still-high rates of population growth. They are trying to grow food on steep hillsides or areas carved out of tropical forests that are highly erodible and of limited fertility, forcing them to move on after only a short time. Environmental writer and consultant Norman Myers calls these people shifted cultivators—a play on the traditional term shifting cultivators—and they can be found in Bolivia, Côte d'Ivoire, Ecuador, Indonesia, Madagascar, Myanmar, Peru, the Philippines, Thailand, and at least a dozen other countries. No reliable totals are available, but Myers estimates the number of shifted cultivators at roughly 250–500 million worldwide, and he foresees a rapid growth in their numbers.[2]

According to U.N. Environment Programme (UNEP) estimates at the beginning of the nineties, some 3.6 billion hectares—nearly a quarter of the earth's land area, or about 70 percent of potentially productive drylands—are affected by desertification. One third of all agricultural land is lightly degraded, half is moderately degraded, and 16 percent strongly or extremely degraded. The annual loss of productive land amounts to some 6–7 million hectares. Some 400 million poor people live in rural, ecologically fragile areas of the developing world characterized by land degradation, water scarcity, and reduced agricultural productivity.[3]

Although no good statistics exist for soil erosion, Myers puts the annual loss of topsoil worldwide at about 26 billion tons, enough to grow some 9 million tons of food—equal to 0.5 percent of global grain production. Erosion removes an estimated 5–10 tons per hectare annually in Africa, Europe, and Australia, 10–20 tons in the Americas, and nearly 30 tons in Asia. Far outweighing the pace of natural soil creation of about 1 ton per year, these rates are clearly unsustainable. Unless remedial action is taken, nearly two thirds of all cropland will perform below its potential in the next four decades.[4]

Land degradation has different sources in various regions. In Africa and Australia, overgrazing is the most important cause of stress; in North America, it is unsustainable agricultural practices; in South America, Asia, and Europe, deforestation is the leading culprit. (See Table 2–1.) The portion of agricultural land affected by soil degradation comes to 65 percent in Africa, 45 percent in South America, 38 percent in Asia, and 25 percent in North America and Europe. Among the soil erosion crisis areas are the Horn of Africa, eastern Iran, large patches of Iraq, the northwestern and northeastern corners of the Indian subcontinent, Central America, the Amazon basin, and several parts of China. In Mexico, for instance, at least 70 percent of agricultural land is affected by soil erosion.[5]

TABLE 2–1. *Causes of Land Degradation, by Region*

Region	Degraded Area	Deforestation	Overgrazing	Croplands	Other
	(billion hectares)	(percent of degraded land)			
Asia	0.75	40	26	27	7
Africa	0.49	14	49	24	13
South America	0.24	41	28	26	5
Europe	0.22	38	23	29	10
North and Central America	0.16	11	24	57	8
Oceania	0.10	12	80	8	0
World	1.96	30	35	28	8

SOURCE: See endnote 5.

About 80 percent of China's soil loss is occurring in 11 poor, interior provinces that are home to a third of the population. If the pace of soil degradation and the outright loss of cropland cannot be offset by a commensurate boost in agricultural productivity, China's future food self-sufficiency may be compromised. This is an issue of concern particularly as that nation's population continues to increase and as growing disposable incomes are boosting per capita demand. The impact of such a trend on food security can hardly be overstated: not only is China itself home to one fifth of the global population, but if China were to become a major food importer, it would squeeze world grain markets. The resulting boost in prices would reverberate through all grain-importing countries.[6]

Norman Myers's "shifted cultivators," along with commercial loggers and cattle ranchers, are the primary agents of deforestation. Deforestation has a number of negative consequences for people's livelihoods and can even threaten their lives. By accelerating soil erosion, altering local hydrological cycles and precipitation patterns, and decreasing the land's ability to retain water during rainy periods, deforestation contributes to reduced agricultural productivity, more severe flooding, and a siltation of rivers and coastlines that may damage fisheries.[7]

A large part of the Indian subcontinent, for example, is highly dependent on the integrity of forests in the Himalayan foothills and other catchment areas in the region. India receives the bulk of its water from just a few monsoon storms each year. Land that is able to absorb water and release it in moderate proportions yields a manageable, appropriate supply; a denuded landscape, however, means devastating floods. The subcontinent now has substantial deficits in this regard. Between half and three quarters of the middle mountain range in Nepal has been deforested in the last 40 years, India has lost 40 percent of its forest cover in the last 30 years, and Pakistan has only a fraction of the forested area that even the government deems necessary. With insufficient forest cover, write M. Abdul Hafiz and Nahid Islam of Bangladesh, "the Himalayas are no longer able to buffer the powerful monsoon because the sponge effect of the soil to absorb a certain amount of water is lost. Hard rain flows more quickly off mountain sides taking...soils that are left unprotected."[8]

The flood-prone areas in India have almost tripled in the last 25 years, now amounting to some 600,000 square kilometers, or almost one fifth the country's total territory. Flooding in India's Ganges Plains alone causes damage of about $1 billion each year to crops, buildings, and public infrastructure. In Bangladesh, even though floods are a regular phenomenon, their severity and frequency has increased: since 1955, the area affected by major floods has almost doubled, to 90,000 square kilometers, or almost two thirds the country's territory. The country used to experience a major flood every four years on average; between 1980 and 1992, however, there were five floods, each imposing unprecedented damage. The 1988 flood submerged 80 percent of the country under water, left 25 million people homeless, and killed tens of thousands.[9]

The consequences of deforestation reach beyond ecological degradation. The loss of woodlands impoverishes and destroys communities of indigenous peoples whose livelihoods, cultures, and sense of self are intrinsically bound up with the forest. Individuals may physically survive, but their communal and spiritual existence is wiped out. Tribal communities are being displaced by commercial logging and by forest clearing for ranching, oil exploration, and other activities, as has happened in the rain forests of Central America, the Amazon basin, India, and other areas.[10]

* * * *

Water, like cropland, is a fundamental resource for human well-being—for food production, health, and economic development. Yet in many countries it is an increasingly scarce resource, under threat of both depletion and pollution. Countries with annual supplies in the range of 1,000–2,000 cubic meters per person are generally regarded as water-stressed, and those with less than 1,000 cubic meters are considered water-scarce. By the beginning of the nineties, 26 countries—home to about 230 million people—were in the water-scarce category. (See Table 2–2.) As water demand grows with population and economic development, their ranks are expected to swell.[11]

Many rivers and aquifers—and not just in countries with acute water scarcity—are overexploited. Excessive withdrawal of river and groundwater leads to land subsidence, intrusion of salt water in coastal areas, and desiccation of lakes. As groundwater is drawn at a rate surpassing natural replenishment, water tables decline. Eventually, the water becomes too costly to continue pumping or too saline for irrigation purposes, or is depleted altogether. Groundwater depletion can permanently reduce the earth's natural capacity to store water. In the United States, more than 4 million hectares—about one fifth of the total irrigated area—are watered by pumping in excess of recharge. Aquifer depletion due to overpumping is occurring in crop-growing areas

TABLE 2–2. *Selected Water-Scarce Countries, 1990 and 2025*

Country	Water Supply 1990	Water Supply 2025
	(cubic meters per person)	
Nigeria	2,660	1,000
Ethiopia	2,360	980
Iran	2,080	960
Peru	1,790	980
Haiti	1,690	960
Somalia	1,510	610
South Africa	1,420	790
Egypt	1,070	620
Rwanda	880	350
Algeria	750	380
Kenya	590	190
Israel	470	310
Jordan	260	80
Libya	160	60
Saudi Arabia	160	50

SOURCE: See endnote 11.

around the globe, including regions of China, India, Mexico, Thailand, northern Africa, and the Middle East.[12]

Another problem is waterlogging and salinization brought about by poor water management. Both reduce crop yields; high salinity can render the land unfit for continued agricultural use. Salt buildup in the soil is thought to reduce crop yields by about 30 percent in Egypt, Pakistan, and the United States. Mexico is losing 1 million tons of grain a year to this problem—enough food for 5 million people. Data are sketchy, but it is estimated that some 25 million hectares—about 10 percent of irrigated cropland worldwide—suffer from salt accumulation, 15 million hectares of which are in developing countries. Each year, an estimated 1–1.5 million hectares are added—in effect offsetting about half the land newly brought under irrigation. Against a backdrop of a declining amount of per capita irrigated land worldwide, this is a worrisome development.[13]

What land degradation is to agriculture and animal husbandry, overfishing is to marine sources of food. Following tremendous growth during the past several decades, the global marine fish catch has stagnated at slightly above 80 million tons a year since the late eighties (with the freshwater catch and aquaculture adding another 20 million tons). The per capita fish catch hit a plateau in the late sixties. The U.N. Food and Agriculture Organization (FAO) in Rome warns that "there is little reason to believe that the global catch can...expand." Moreover, the composition of the catch has shifted from higher-valued to lower-valued species, and from mature to younger fish. To the extent that fish are caught at earlier stages in their development, stocks are less able to recover fully. Excessive harvesting can diminish stocks to such an extent that reproduction cannot keep up with natural loss.[14]

In March 1995, FAO announced that 70 percent of the world's fish stocks were at some stage of deterioration through overfishing. All 15 major fishing areas are close to reaching or have already exceeded their natural limits. The catch in all but 2 areas has fallen. Among the most affected areas are several parts of the Atlantic Ocean and the east-central Pacific, where the catch has shrunk by more than 30 percent. The Newfoundland cod fishery off Canada is now severely depleted. In July 1992, the Canadian government was forced to declare an indefinite moratorium on cod fishing, only six years after a record harvest was brought in.[15]

As these traditional fisheries are reaching their limits, fishing fleets are increasingly scouring the deep seas. But the fish that inhabit these waters are much more vulnerable because they reproduce more slowly. Stocks of orange roughy around New Zealand, for instance, have already collapsed. Relatively little is known about deep-sea ecology, but there is concern that the depletion of fish there could have unforeseen repercussions, such as a disruption of the oceanic food chain.[16]

Overfishing—including the use of trawlers and wasteful fishing practices—is one reason for the severe stress on fisheries. Pollution and degradation of habitat and spawning grounds are other reasons. The ecosystems of the coastal areas (where 90 percent of the global fish catch takes place) are under severe assault from industry, settlements, tourism, and other factors. Many coastal wetlands, estuaries, bays, mangrove forests, and coral reefs that are critical to the regeneration of stocks are at risk. A 1995 study by the Washington-based World Resources Institute found that half the world's coastlines are at medium or high risk, with those in Europe and Asia topping the list.[17]

The issue is not just one of future marine food supplies for an expanding human population—as important as that is. Stagnation and a potential future decline in the fish catch have important ramifications for the coastal communities whose livelihoods are based on fishing. Already the ranks of fishers—an estimated 15–21 million persons worldwide—are thinning. More job loss seems inevitable, as the global fishing industry has twice the capacity it needs to prowl the world's oceans. And beyond the fishers are another 180 million people whose livelihoods depend on fish-related industries.[18]

A 1995 study found that 125 of the commercial fish

species could bounce back if the catch were reduced sufficiently. Norway did impose strict limits on the cod catch in its territorial waters, following overfishing in the eighties; stocks recovered enough to permit at least a reduced level of fishing. Regulations to limit the catch would allow fishing to become more sustainable; a free-for-all, by contrast, yields a temporary increase in catch but, by pushing against the limits, risks eventual collapse. Yet political leaders are pressured to maintain oversized capacities.[19]

Large-scale fishing operations are geared to commercial markets in which the highest bidders rule. As supplies tighten and consumers in rich countries absorb a growing share of the global catch—75 percent of all fish caught go to industrial countries—prices rise. Poorer people are losing access to fish at affordable prices, even though many rely on it as their primary source of animal protein. (In Africa, fish supplies 19 percent of the protein in people's diets, and in Asia, 29 percent.) In Kerala, India, for example, per capita fish consumption was cut in half between 1971 and 1980 as the state's fishers turned to commercial markets and prices skyrocketed. As Deborah Cramer, writing in the *Atlantic Monthly*, put it: "International markets are siphoning fish away from those who need it to those who merely want it."[20]

Taken together, these trends of increasing land degradation, water scarcity, and overfishing suggest great difficulty in providing for the food needs of a global population that is growing fast. If the planet's social and economic systems were geared to sharing the planetary product equitably, there would be enough food to go around—at least for the time being and perhaps even for a long time to come. But that is not the case, and any scarcities on a global scale will affect most those who already suffer from inadequate food supplies: scarcities will translate into higher prices and push them out of the market. The ranks of the food-insecure everywhere may well rise unless adequate measures are taken to address the land, water, and fishery issues of our time.

★ ★ ★ ★

The already observable degradation of the earth's soil, water, and fishery resources, which will have varying effects on national and global security, is likely to be compounded by atmospheric ozone depletion and climate change—two phenomena that scientists have long hypothesized about and that governments around the world have only begun to address seriously.

The depletion of the ozone layer that protects the environment and people from harmful ultraviolet (UV-B) radiation continues to worsen year after year, and has in fact proceeded faster than scientists expected. The extent of depletion varies seasonally and according to latitude, and exposure to increased radiation changes with cloud cover. The largest ozone hole ever measured—covering 10 million square kilometers, about the size of Europe—appeared over Antarctica in mid-1995. Ozone levels over the region declined 10 percent from 1994, the previous record low. But more inhabited regions, including parts of Europe, North America, Australia, and southern Latin America, are now also subject to significant increases in radiation. Above Europe and North America, for instance, ozone levels have declined by 10 percent since the late fifties.[21]

A 1-percent depletion of the ozone layer translates into a 2-percent increase in the radiation dose. Even small increases in radiation are likely to have important ramifications for human health and for plant and animal life. A sustained 10-percent decrease in the average ozone concentration could lead to some 250,000 additional non-melanoma skin cancers each year. Increased ultraviolet radiation can also weaken the human immune system, leaving people more susceptible to infectious diseases.[22]

Enhanced UV-B radiation further poses a threat to agricultural harvests and to marine fisheries. Simulations of increased radiation have shown that crop yields may drop 5 percent for wheat and as much as 90 percent for squash; one quarter of soybean yields may be lost. In the marine environment, phytoplankton is highly susceptible to increased radiation, and has apparently already been adversely affected in the southern reaches of the oceans. Moving up the food chain, the effects are likely to be seen in weakened fish stocks and reduced fish catches.[23]

Due to the 1987 Montreal Protocol and several follow-on agreements to phase out the major chemicals implicated in ozone destruction, the production of chlorofluorocarbons (CFCs) fell by 77 percent between 1988 and 1995. Growth in the concentration of ozone-depleting chemicals in the atmosphere is now slowing. But given the large amounts of CFCs and other ozone-depleting chemicals that have already been emitted and their long lifetimes in the atmosphere, ozone depletion will get worse before it improves. The maximum depletion—and hence the greatest increase in radiation—is expected to occur between 1997 and 1999. It will be at least 2050 before chlorine levels return to where they were in the late seventies, when the ozone hole was first discovered.[24]

Not nearly as much progress has been achieved with regard to averting climate change, even though the scientific consensus that global warming is indeed under way has become stronger and stronger in recent years. Critics of the "greenhouse effect" point to weaknesses in computer models used to simulate and predict climatic change, and to inevitable scientific uncertainties. But these should give rise to precaution rather than a dismissal of the notion of global warming.

The accumulation of carbon dioxide and other gases in the atmosphere that trap heat is believed to be responsible for potentially catastrophic alterations in the global climate. The Intergovernmental Panel on Climate Change (IPCC), a body of scientific experts set up by the United Nations, stated in November 1995 that "the balance of evidence suggests that there is a discernible human influence on global climate." The group projected an average increase in global temperatures of 1.5–6.3 degrees Celsius by 2100 if no action is taken.[25]

Changing precipitation patterns, shifting vegetation zones, and rising sea levels caused by global warming threaten to disrupt a wide range of human and natural systems. These would likely disrupt crop harvests, inundate heavily populated low-lying coastal areas, upset human settlement patterns, threaten estuaries and coastal aquifers with intruding salt water, and undermine biological diversity. A hotter climate could trigger an increase in heat waves, hurricanes, floods, droughts, fires, and pest outbreaks in some regions; more extreme climates in desert zones; a rise in the number of heat-related deaths and illnesses; and expansion in the reach of vectorborne infectious diseases such as malaria, yellow fever, dengue fever, and viral encephalitis. Many ecosystems cannot adapt to a rapid change in climatic conditions and zones; entire forest types could disappear.[26]

Global warming could cause sea levels to rise anywhere from 15 to 94 centimeters during the next century (and more thereafter), with a current best estimate of about 50 centimeters. Up to 118 million people in coastal areas could be put at risk. Several island nations might disappear entirely, and not surprisingly their governments have become leaders in the effort to formulate a global policy to avert further warming.[27]

After the island nations, Bangladesh will be the most severely affected. The currently projected maximum sea level rise would inundate about 25,000 square kilometers of the country—17 percent of its territory. The effect of rising seas would be compounded by the ongoing process of land subsidence and by the likely large volumes of water that would be released as Himalayan glaciers melt due to warmer temperatures. Bangladesh would also face more frequent and more intense tropical storms. Greater differences in temperature between sea and land would make the monsoon system more powerful and violent. The coastal part of Bangladesh could be permanently submerged, and virtually the entire country would be subject to repeated massive flooding and natural disasters. The sustenance of some of the poorest peasants cultivating the *char*—shifting sandbanks along the coast and sides of rivers—could literally disappear beneath the waves.[28]

Severe repercussions would be felt elsewhere as well. Egypt, for example, could lose 15–19 percent of its habitable land within a few decades, displacing a similar percentage of its population. Other river deltas and coastal areas around the globe affected by global warming include the Yangtze, Mekong, and Indus in Asia; the Tigris and Euphrates in the Middle East; the Zambezi, Niger, and Senegal in Africa; the Orinoco, Amazon, and La Plata in South America; the Mississippi in North America; and the Rhine and Rhone in Europe.[29]

The low-lying areas most at risk are precisely the places with some of the densest human settlements and the most intensive agriculture. All in all, UNEP anticipates that sea level rise, along with amplified tidal waves and storm surges, could eventually threaten some 5 million square kilometers of coastal areas worldwide. Though accounting for only 3 percent of the world's total land, this area encompasses one third of all croplands and is home to more than a billion people. The IPCC points out, for example, that almost 10 percent of the world's rice production, feeding more than 200 million people, takes place in areas of Asia that are considered vulnerable to sea level rise.[30]

Global warming's impact on agriculture—through rising seas, higher or more variable temperatures, more frequent droughts, and changes in precipitation patterns—is indeed a major concern. Higher temperatures mean greater evaporative losses and hence faster desiccation of soils. The effects would be highly uneven, with some areas benefiting from changes in temperature and alterations in the hydrological cycle, but others that now receive plentiful rainfall becoming substantially drier.

Given current water shortages, agriculture in arid and semiarid areas is particularly vulnerable to climate change. According to Norman Myers, "Regions that appear to be at greatest risk of extreme climatic dislocations for agriculture are often those where marginal environments sometimes make agriculture an insecure enterprise already: the Sahel, southern Africa, the Indian subcontinent, eastern Brazil, and Mexico." But some important food baskets, such as the U.S. Midwest, are also projected to receive less rainfall on average and to experience drought conditions more frequently. The mid-1996 drought in the U.S. Great Plains, the worst since the thirties, is a clear warning of what climate change may bring.[31]

Studies suggest that Egypt's corn yield may drop by one fifth and its wheat yield by one third, while Mexico's rain-fed maize crop may be reduced by as much as 40 percent by global warming. Myers points out that India's wheatlands in Harayan, Rajasthan, and Punjab, "vital to the nation's capacity to feed itself, are at the northern edge of their range. They could not tolerate a temperature increase of more than 2 degrees Fahrenheit, yet global warming in India may well exceed this amount within just a few decades. Nor could the wheatlands 'migrate' northwards, since they would run up against the Himalayas."[32]

* * * *

The human security implications of degraded or depleted lands, forests, and marine ecosystems, intensified by ozone depletion and climate change, are many: heightened droughts and increasing food insecurity; reduced crop yields that force peasants to look for work in already crowded cities; environmental refugees from coastal areas seeking new homes and livelihoods elsewhere, possibly clashing with unwelcoming host communities; the disruption of local, regional, and possibly national economies; and the soaring costs of coping with the dislocations. To some extent, these effects simply mean a continuation of already observable events, though on a much larger scale.

The social, economic, and political repercussions are stark. As competition over increasingly depleted and degraded resources increases and as migrant and refugee streams swell, the environment could become more important in triggering or aggravating conflicts.

3

Conflict Over the Environment

> Our land is being polluted, our water is being polluted, the air we breathe is being polluted with dangerous chemicals that are slowly killing us and destroying our land for future generations. Better that we die fighting than to be slowly poisoned.[1]
>
> Bougainville Revolutionary Army, 1990

This dramatic declaration was issued by a group that since 1988 has fought a ferocious war of secession from Papua New Guinea. The conflict was triggered largely by the environmental devastation caused by a huge copper mine on the island of Bougainville that operated to the virtually exclusive benefit of the central government and foreign shareholders.

Bougainville is a somewhat unusual case—both because the environment is clearly identified as the primary, immediate cause of conflict and because the situation involved substantial violence. Environmental degradation certainly does not always lead to conflict, particularly violent conflict. Among sovereign states, most disputes over environmental matters are typically dealt with in the diplomatic, bureaucratic, or other political realm rather than by force of arms. Although governments may yet decide to resort to force in some cases—conflicts over scarce water resources being the prime example—these tend to involve countries that are locked in antagonistic relationships with their neighbors and lack adequate structures to address shared problems.

Environmentally induced conflict is more likely within individual countries. This is because the needs and interests of contending groups tied closely to land and environmental resources—peasants, nomads, pastoralists, ranchers, resource extractors—often remain unreconciled. These contending interests are typically bound up with issues of ethnicity and economic development, of subsistence versus commercial operations. Governance structures are often incapable of adjudicating conflicting interests, favoring instead one group over another.

★ ★ ★ ★

It may be the social, economic, and political repercussions of environmental change—rather than the change itself—that are the most important determinants of conflict over the environment. Communities directly and greatly affected by soil erosion or water scarcity, for example, may see their agricultural potential diminished, parts of their populations uprooted, and their traditional ways of life disintegrating. The effects will be particularly pronounced if economic decline, political instability, or ethnic polarization are already present.[2]

The ability of different societies to cope with the effects of environmental stress varies considerably. First, technical "ingenuity"—the capacity to devise scientific and engineering responses to deal with contamination problems or to circumvent scarcities—is much more highly developed in industrial countries. (Of course, this point accepts the western view of the preeminent role of technology, ignoring the growing understanding that many environmental challenges are, in the final analysis, not of a technical but a social nature.)

Second, countries with broadly diversified, vibrant economies have a greater ability to withstand environmental stress. Countries highly dependent on agriculture and primary commodities—principally those in sub-Saharan Africa, South Asia, and Central America—face a much greater threat to their stability and integrity from environmental degradation.[3]

Third, a country will also be better able to face environmental stress if its social resilience—the strength and cohesion of the communal fabric—is strong. There are enormous differences among countries in this regard, but the fabric of many is now subject to intense erosive forces. (See Chapter 4.)

Furthermore, the ability of political and civic leaders to fashion adequate policy responses is a key asset, influenced by the authority and legitimacy that a government can marshal. All too often, however, instead of addressing grievances, political leaders seek to exploit pressures and divisions for their own benefit, greatly aggravating the fissures within a society.[4]

It is only in the context of complex social, economic, and political processes that the impact of environmental decline—whether it contributes to conflict or cooperation—can be evaluated. On the whole, the countries least able to withstand any additional stress are those that already suffer from the classical signs of underdevelopment (such as poverty, unequal land distribution, rapid population growth, or a huge foreign debt), that manifest deep ethnic tensions or social and other cleavages within their populations, and that have nonrepresentative governance structures.[5]

Several countries in Africa, Asia, and Latin America are clearly in this general category, and the examples in this chapter are drawn from them. Industrial countries are certainly not immune to the deleterious and destabilizing effects of environmental degradation, but they have more of a buffer to cope with adverse impacts and greater leeway to finance remedial measures.

★ ★ ★ ★

Within countries, it is minority groups, indigenous peoples, and other vulnerable and impoverished communities such as subsistence peasants or nomadic tribes that often bear the brunt of adverse environmental transformation—particularly that triggered by resource extractive operations. Their capacity to resist and defend their interests is very weak. These groups not only have to eke an existence out of marginal, typically arid or semi-arid lands, they are also socially and politically marginalized. Often powerless to struggle for the preservation of natural systems that their livelihood and very survival depends on, they may face a future of further suffering and displacement. This is particularly the case if they belong to communities steeped in a culture of fatalism.[6]

But affected groups do not always simply submit to this fate, as events in Bougainville demonstrate. The Panguna copper mine on the island has had a devastating effect on the local environment and the traditional way of life of its people. Operations at the mine, one of the world's largest and the mainstay of the Papua New Guinea economy, began in 1972. Over the years, subsistence agriculture and traditional hunting and gathering suffered severely from the large-scale strip-mining operations. Mine tailings and pollutants covered vast areas of land, decimated harvests of cash and food crops including cocoa and bananas, and blocked and contaminated rivers, leading to depleted fish stocks. All in all, about one fifth of the island's total land area has been damaged. With additional minerals prospecting under way, local people began to fear that an even greater portion of the territory would eventually be destroyed by mining operations.[7]

Bougainvilleans received few of the economic benefits of all this environmental destruction. The bulk of the profits from the mine went to the national government and foreign shareholders. Royalty payments to local landholders amounted to only 0.2 percent of the cash revenue of the mine, and compensation payments—for land leased and damage wrought—were seen as inadequate. The concerns and demands of Bougainville's inhabitants were ignored by both the central government and the British/Australian-owned copper company. Discontent continued to build, with local landowners in the forefront of opposition to the mine. By late 1988, a sabotage campaign had begun; it soon developed into the guerrilla war that closed the mine in May 1989. Papua New Guinea lost 40 percent of its foreign-exchange earnings.[8]

The population of Bougainville is ethnically distinct from that of other parts of Papua New Guinea. Although secessionist aspirations existed before the Panguna mine opened, Volker Böge of the University of Hamburg observes that "it was only after demands for environmental compensation had been refused that ideas of secession came to the forefront." Bougainville declared its independence in May 1990, an act not recognized by any other state. The conflict continued, bringing both fighting and intermittent cease-fires and talks. It remains unresolved.[9]

Indigenous groups in Nigeria's delta region are facing a threat quite similar to that of Bougainvilleans—the massive despoiling of their environment by resource extraction operations run by multinational corporations. More than 90 percent of Nigerian oil is produced in the Niger delta, home to indigenous groups of the Urhobo, Ijaw, Isoko, Kalabari, Ndokwa, Itshekiri, and Ogoni. Traditionally, they lived from fishing, agriculture, and palm oil production, but petroleum production is destroying the basis of their livelihoods. Among these groups, the Ogoni—numbering about a half-million people—have been most active in demanding environmental assessments, cleanup, and a fairer share of the economic benefits of oil production, as well as most adept at generating international support for their cause.[10]

Frequent oil spills from antiquated pipelines, the flaring of natural gas, leaks from unlined toxic waste pits, and generally poor practices have exacted a heavy toll on soil, water, air, and human health. Formerly lush agricultural land is now covered by oil slicks, and much vegetation and wildlife has been destroyed. Many Ogoni suffer from respiratory diseases and cancer, and birth defects are frequent. Like Bougainvilleans, the Ogoni suffer the negative consequences of resource extraction without receiving a commensurate share of the benefits.[11]

Today, the area remains impoverished. It has no electricity or plumbing, and the few roads primarily provide access for the oil industry. Schools are more often closed than open, and the sole hospital in the region remains unfinished.[12]

In early 1992, the Movement for the Survival of the Ogoni People (MOSOP) was created. It led a peaceful campaign against what it called "environmental terrorism." Shell is by far the leading oil company in Nigeria. Although the company operates in about 100 countries, fully 40 percent of all Shell oil spills worldwide have occurred in Nigeria. MOSOP drew up a Bill of Rights for the Ogoni and called for true representation in the Nigerian Federation. But the military government, described by a Nigerian professor as the "patron saint" of the oil companies, was more concerned with keeping Shell and other multinationals happy than with taking care of the local population: oil income accounts for 80 percent of government revenues and is probably lining the pockets of the regime.[13]

The military government conducted a violent campaign to intimidate the protesters. Repression escalated in January 1993, when soldiers opened fire on demonstrators. Villagers have been subjected to killings, torture, and rape. The military is also widely believed to have instigated attacks on the Ogoni by the neighboring Andoni ethnic group. The government and Shell afterwards blamed the violence on ethnic antagonisms. All in all, at least 27 Ogoni villages have been destroyed, some 2,000 people killed, and 80,000 uprooted. The leaders of MOSOP were either detained or forced to go into hiding. In May 1994, Ken Saro-Wiwa, MOSOP's well-known spokesperson, was imprisoned and held on contrived murder charges; ignoring worldwide protests, the military dictatorship executed him and eight other Ogoni activists in November 1995. The Ogoni homeland is effectively under military occupation.[14]

Shell ceased operating in Ogoni territory in 1993, but it remains the major oil producer in the country. Indeed, immediately after the executions, Shell gave the go-ahead for a $4-billion natural gas project in partnership with the regime. A variety of government and international sanctions were imposed on Nigeria, but no government has taken the one step that would make a real difference: an embargo on Nigerian oil. In the capitals of the United

States and Western Europe, which together consume virtually all of Nigeria's oil exports, the profit interests of a large multinational company apparently count more than the survival interests of an entire ethnic group.[15]

The Bougainville and Ogoni examples are unusual only in that the former escalated into a full-scale war and the latter has received considerable media and public attention. Minority populations and indigenous peoples around the globe are facing massive degradation of their environments that threatens to irreversibly alter, indeed destroy, their ways of life and cultures.

In many cases, central governments promote unsustainable mining, logging, ranching, and other projects. These may be designed to provide revenue to service foreign debt, but they typically help prop up unrepresentative, sometimes repressive regimes and enrich national elites and foreign corporations, with few benefits accruing to those whose lands are devastated. Among current or recent examples are conflicts about oil drilling in Myanmar, Ecuador, and Peru, about mining in Indonesia's Irian Jaya province, and about mining and logging in Suriname. Resistance to these operations usually brings swift repression and human rights violations.[16]

★ ★ ★ ★

Throughout human history, societies have clashed over access to and control of ostensibly renewable resources such as water and farmland and over nonrenewable ones such as oil and minerals. But the competition over resources that was once a simple zero-sum game—to the victor belonged the spoils—is being transformed by environmental scarcities. Although countering the degradation requires cooperation among countries that share watersheds and other ecosystems, the immediate effect is that competition between them intensifies, particularly as populations and development needs continue to grow.

An estimated 40 percent of the world's population depends for drinking water, irrigation, or hydropower on the 214 major river systems shared by two or more countries; 12 of these waterways are shared by five or more nations. For some countries, almost the entire flow of surface water originates beyond their own borders—more than 90 percent of Botswana's, Egypt's, and Hungary's, for example. Disputes between upstream and downstream riparians over water use and quality simmer in virtually all parts of the world. These involve reduced water flow and siltation because of dams, water diversion for irrigation, industrial and agrochemical pollution, salinization of streams due to unsound irrigation practices, and floods aggravated by deforestation and soil erosion. (See Table 3–1.)[17]

Peter Gleick of the Pacific Institute for Studies in Development, Environment and Security notes that "not all water resources disputes will lead to violent conflict; indeed most lead to negotiations, discussions, and nonviolent resolutions." In some parts of the world, river commissions with representatives of riparian countries provide a forum in which disputes can be addressed adequately; elsewhere, however, generally adversarial relationships among riparian states make for a much greater challenge. "There is growing evidence," says Gleick, "that existing international water law may be unable to handle the strains of ongoing and future problems."[18]

The Middle East and southern Asia are the areas of highest concern. Water scarcity and water allocation have already played an important role in past Arab-Israeli wars, and now pose a tremendous challenge in the peacemaking process. About 40 percent of the groundwater

TABLE 3–1. *International Water Disputes of Varying Intensity, Eighties and Nineties*

Body of Water	Countries Involved	Subject of Dispute
Nile	Egypt, Ethiopia, Sudan	Water diversion, flooding, siltation
Euphrates	Iraq, Syria, Turkey	Reduced water flow, salinization
Jordan, Yarmuk, Litani	Israel, Jordan, Syria, Lebanon	Water flow, diversion
West Bank	Israel, Palestinians	Water allocation and aquifer water rights
Indus, Sutlei	India, Pakistan	Irrigation
Brahmaputra, Ganges	Bangladesh, India	Water flow, siltation, flooding
Salween/ Nu Jiang	Burma, China	Siltation, flooding
Mekong	Cambodia, Laos, Thailand, Vietnam, China	Water flow, flooding
Aral Sea	Kazakstan, Uzbekistan, Turkmenistan	Repercussions of shrinking sea, water scarcity, salinization
Paraná	Argentina, Brazil	Dam, land inundation
Lauca	Bolivia, Chile	Dam, salinization
Rio Grande/ Colorado	Mexico, United States	Water flow, salinization, pollution
Rhine	France, Germany, Netherlands, Switzerland	Industrial pollution
Szamos	Hungary, Romania	Industrial pollution
Danube	Hungary, Slovakia	Dam, flooding

SOURCE: See endnote 17.

Israel depends on originates in occupied territories, and Palestinian water drilling and pumping have been under severe restrictions since 1967. Palestinian wells have run dry or become saline, and their irrigated agricultural areas declined from 27 percent to 4 percent of total arable land. It remains to be seen whether Israel is prepared to allow a meaningful change in the current, highly unequal water allocation, particularly following the change of government in mid-1996. Without it, there may not be a lasting peace.[19]

Across the Red Sea, the Nile waters are in contention. Ethiopian plans to expand greatly the diversion of Nile water for irrigation and hydropower purposes will bring that country into direct conflict with Egypt, for whom the Nile is a lifeline. Similarly, conflicting water use plans for the Euphrates River have created tension among Turkey, Syria, and Iraq. A Syrian dam project brought Syria and Iraq to the brink of war in 1974. Turkey's Grand Anatolia Project, involving the construction of a series of dams, in effect provides Ankara with a "water weapon": in 1992, it threatened to restrict the flow of water into Syria if Damascus persisted in supporting Kurdish rebels in Turkey.[20]

Violent conflict over water resources is not highly likely in cases where the downstream country has less power than the upstream country, even though the former may suffer substantial social and economic insecurity. There is perhaps no better illustration of this than the India-Bangladesh relationship. The two countries share 54 rivers between them, including the Ganges. But since the rivers originate beyond its borders, Bangladesh has virtually no control over the waterways that can deliver life or death.[21]

Bangladesh is marked by what native writers M. Abdul Hafiz and Nahid Islam call the "paradox of water": during the monsoon season, it often receives too much water, resulting in widespread flooding; but in the December-April dry season, it receives too little, resulting in serious drought conditions. Some 18–19 million Bangladeshis are affected by flooding each year, and more than 1 million people have been killed since 1961. Yet floods are a "necessary evil," as *New York Times* reporter Sanjoy Hazarika puts it: necessary to replenish soil fertility.[22]

Heavy diversion of water by India for irrigation purposes exacerbates these natural conditions. In 1975, India completed the Farakka Barrage on the Ganges river. The altered flow of the Ganges has made the riverbed less able to handle the heavy monsoon rain and has increased the vulnerability to flooding. During the dry season, the flow of the Ganges (as measured at the border) is now one seventh its pre-Farakka level, and the river no longer reaches the Bay of Bengal. Consequently, saline water is intruding into the western portion of the river delta in Bangladesh, contaminating drinking water supplies and destroying trees and farmland. Groundwater levels in many areas have dropped by more than 3 meters.[23]

In the affected area—some 26,000 square kilometers—crops, fisheries, and mangrove forests are being damaged. Their economic base undermined, many rice growers in the river plains area have abandoned their fields and migrated to the shantytowns of Dhaka. By 1995, India's water withdrawals were estimated to reduce Bangladeshi food production by more than 3 million tons annually. Crop losses cost Bangladesh some $1.25 billion a year.[24]

India is withdrawing water not only from the Ganges but also from the Brahmaputra, the Teesta, and several other, smaller rivers. Several interim agreements over water sharing between India and Bangladesh were negotiated in the seventies and eighties, but the two countries have been deadlocked over the issue for the past decade. India, as the upstream riparian and far stronger economically and militarily, is able to regulate the water flow according to its own designs. Although Dhaka presses the issue in a variety of international forums, India has consistently rejected any third-party mediation.[25]

Water disputes take place not just between states, but also within them. One form of this conflict is seen between different regions of a country—typically, regions that are more arid or have already exhausted their own supplies claiming the water resources of more amply endowed areas. China, India, Mexico, the United States, Spain, and the former Soviet central Asian republics are among the countries in which growing water scarcities and regional disparities either have caused or could cause conflict.[26]

In India, disputes over sharing the waters of the Cauvery River between the states of Karnataka and Tamil Nadu triggered strikes, riots, arson, and looting in 1991 that left several people dead and led to the temporary exodus of 100,000 Tamils from Karnataka. In Mexico, the states of Nuevo Leon and Tamaulipas had a legal and political feud for months before agreeing on a water-sharing plan for the San Juan river.[27]

In China, water is already being siphoned away from agriculture in the Beijing area to serve burgeoning urban demand. Central planners are promoting several large-scale river diversions in different parts of the country. China expert and geographer Vaclav Smil of the University of Manitoba predicts that some of these projects could trigger major interprovincial conflict and pose a rising challenge for the central government. Among the large planned diversions is that of the Huang

He (Yellow River). Shanxi province, with chronic and severe water shortages, would be the beneficiary, but in dry seasons, northern Henan and northwestern Shandong could face what Smil calls unprecedented and crippling economic losses and prolonged shortages of drinking water for close to 50 million inhabitants.[28]

The much larger diversion scheme planned for the Chang Jiang—to northern Jiangsu, Shandong, Hebei and Tianjin—has triggered concern in eastern Jiangsu and Shanghai about saltwater intrusion into the river's estuary. The waters of the Chang Jiang are also at issue in the construction of the controversial Three Gorges dam. Sichuan, Jiangsu, and Shanghai provinces have a variety of concerns, including the expected extensive flooding of farmland, and the likely detrimental effects on the river delta, wetlands, and coastline, as well as on one of the most important breeding grounds of commercial fish stocks.[29]

The second type of internal conflict surrounding water projects concerns the severe impacts on local populations and the fact that they are typically excluded from the decision making that affects their homes, livelihoods, and way of life. Large-scale schemes such as irrigation and hydroelectric facilities, says Peter Gleick, "often lead to the displacement of large local populations, adverse impacts on downstream water users, changes in control of local resources, and economic dislocations. These impacts may, in turn, lead to disputes among ethnic or economic groups, between urban and rural populations, and across borders."[30]

A study by the International Rivers Network found that the construction or expansion of 604 dams in 93 countries displaced at least 10 million people during 1948–93, most of whom received no compensation or rehabilitation support. This is by no means a complete accounting, and the ranks of the displaced are continuing to swell with additional projects. A 1994 World Bank study put the current displacement toll of dams in developing countries at more than 4 million a year.[31]

Two of the most controversial schemes are Three Gorges in China and Sardar-Sarovar Narmada in India. The Three Gorges dam on the Yangtze is intended to be the world's largest hydroelectrical plant; construction started in 1994 and is expected to continue until 2008. It would uproot some 1.4 million people and flood 11,000 hectares of fertile farmland. By the end of 1995, some 25,000 people had already been relocated, some forcibly.[32]

The Sardar-Sarovar dam project in the Narmada river valley is the keystone of a gigantic dam-building and irrigation project. With crucial financial support from the World Bank, the project was begun in 1987. But its viability has been questioned, and its likely devastating impacts triggered intense opposition both from affected communities in India and from large numbers of nongovernmental organizations around the world.

Close to 100,000 hectares of arable land would be flooded by the dam's reservoir or lost to irrigation canals and other related infrastructure, and another 14,000 hectares of forested areas would also be submerged. The densely populated agricultural land above the dam site would be exposed to greater flooding due to silt buildup in the Narmada, while arable land in the delta may be exposed to saltwater intrusion, and fisheries downstream from the dam would be decimated. It is also expected that the reservoir and irrigation canals would contribute to the spread of diseases such as malaria, filaria, and schistosomiasis.[33]

An estimated 240,000–320,000 people will be displaced by the project—losing not only their homes as some 237 villages are submerged, but also their livelihoods. The Indian government has offered to resettle those affected, in the largest such undertaking ever attempted in India. But those already relocated have come into conflict with host populations and have received land of inferior quality. Many displaced people may end up in urban slums because they do not have officially recognized land-rights documents. And resettlement cannot, in any event, redress the loss of ancestral lands, pilgrimage sites, and religious monuments, nor the associated cultural and spiritual alienation.[34]

As is the case so often with megaprojects, the costs and benefits of Sardar-Sarovar would be distributed very inequitably. The project will primarily benefit a small number of already wealthy farmers in central Gujarat who cultivate water-intensive cash crops such as sugarcane, oil seeds, and cotton for export.[35]

The consent of the affected population was never sought, but opponents formed the Movement to Save the Narmada (Narmada Bachao Andolan, or NBA) and launched a highly effective campaign of civil disobedience and mobilization of world opinion. The government and project operators have resorted to various forms of police repression, but their tactics could not prevent the NBA from scoring some important successes: the World Bank was compelled to initiate an independent evaluation commission (resulting in the highly critical Morse Report in June 1992) and finally to pull out of the project in March 1993. Japan also withdrew its funding.[36]

The Indian government insists it is going ahead with the project—some construction continues and a small part of the Narmada valley has been submerged—but the project operators now face a severe financial crisis and are trying to bring in private investors. The government of Madhya Pradesh (where the reservoir is to be located) announced it was seeking a reduction in the height of the dam because it could not resettle the huge number of people who would otherwise be displaced. The NBA, meanwhile, filed a case with India's Supreme Court, even as police harassment continued.[37]

★ ★ ★ ★

Land degradation is another key category of environmentally induced conflicts. Particularly in arid and semi-arid zones such as the Sahel, land degradation exerts strong pressure on peasants, pastoralists, and nomads, bringing them into heightened conflict against each other and against encroachment by commercial interests and central governments. Land degradation can have explosive consequences: together with unequal distribution of and contested access to land, it was a key factor in the 1969 war between El Salvador and Honduras and the 1977-78 Ogaden war between Somalia and Ethiopia.[38]

Explanations of land degradation and related conflicts often focus on the simple arithmetic of carrying capacity: rising numbers of people and livestock overwhelm available resources. But while population growth certainly pushes cultivators to rely on drier and more easily degraded soils, other dynamics—changing land rights patterns and commercialization—are at play as well, and they are important in considering both the origins of conflicts and possible avenues for their settlement.

In response to burgeoning demand for meat and other animal products, particularly in the wealthy oil states of the Persian Gulf, parts of the Sahel have seen a shift from subsistence to commercial herding. With it came a greater emphasis on cattle and sheep, which are not as well adapted to the environment as goats and camels are. Another aspect of the shift was the move from communal to individual ownership of livestock, and hence the decline of traditional forms of control over economic activity. On the Haud plateau of the Ogaden region, for example, numerous new water holes were drilled, permitting a dramatic increase in livestock grazing. Grazing in this ecologically delicate area had traditionally been limited by the amount of rainwater, and nomadic or seminomadic life-styles were highly adapted to the conditions. But with the expansion of grazing, Astri Suhrke of the Chr. Michelsen Institute in Bergen, Norway, observes, "the ecological equilibrium of the Haud quickly disintegrated."[39]

Additional pressure on fragile ecosystems came in the form of increased cash cropping. As cultivation for export markets expanded, vegetation was cleared on a large scale, particularly after World War II. But the massive clearing of vegetation, together with the effects of overgrazing, had fatal consequences. Denuded and depleted, the soil was unable to retain water sufficiently. When the rains failed, disastrous droughts hit the region in the seventies and mid-eighties.[40]

The Sudan provides an instructive illustration of the consequences. In the late sixties, northern Sudanese elites—primarily an urban Arabic trading class known as the Jellaba—pushed ahead with the expansion of large-scale mechanized farming schemes; initially this took place in the north, but from the late seventies on it occurred primarily in southern Sudan. Backed by the central government and underwritten by the World Bank, these projects caused tremendous social and environmental stress and were a key factor in the turmoil that continues to plague Sudan.

The total area under licensed, large-scale mechanized schemes increased tenfold between 1968 and 1986, to about 4–5 million hectares. This area alone exceeded the land in traditional rain-fed agriculture that provided livelihoods for 2–3 million small peasant farmers. Including areas farmed illegally, mechanized farming covered perhaps 10 million hectares.[41]

The projects devoured large tracts of traditional farm and grazing land, forests, and water points, displacing millions of small producers. Some lost their land outright in expropriations; others were forced out indirectly by declining soil quality, the blockage of traditional herding routes, the increasing scarcity of grazing areas, and other effects of the spread of tractors. The upshot was that traditional agropastoralism began to collapse in large swaths of the country; the ranks of those who were forced to migrate swelled by about 4.5 million between 1978 and 1984.[42]

The environmental effects of mechanized farming were as devastating as those in the social sphere. About 95 percent of the forests in eastern Sudan were cut down. Sudan's fragile soils were rapidly exhausted. According to Mohamed Suliman of the Institute for African Alternatives in London, in some areas the land was depleted in three to four years, with yields of sorghum, millet, and groundnuts falling by up to 80 percent. Some 17 million hectares—half of all arable land in northern Sudan—have been lost to soil erosion.[43]

Sudan's growing assimilation into the world market was an important driving force of the Jellaba's destructive exploitation of resources. Loan conditionalities imposed by the International Monetary Fund (IMF) and the World Bank were instrumental in reorienting agriculture in the mid-seventies from satisfying domestic needs to serving export markets. Suliman argues that this change explains in part why the Sudanese were hurt more by the 1983–85 drought, which resulted in widespread famine and a quarter-million deaths, than by the more severe 1972–75 drought.[44]

As mechanized farming inexorably pushed southward while the economic benefits continued to head north, the local population—the Nuba, Dinka, Ingessana, and other black African tribes—regarded this as a hostile incursion. Many joined the Sudan People's Liberation Army (SPLA) of insurgent forces. The dispute over mechanized farming was a key contributing factor to the renewed confrontation and war between north and south Sudan.

The mechanized schemes, along with associated projects, became prime targets of attack. Among them was the planned Jonglei Canal, which was supposed to have delivered water to the Malakal cotton-growing project in Upper Nile province and, by drying out the southern Sudd swamps, would have opened that area for mechanized farming. The southward march of the tractors was stopped when war erupted in 1983. Resistance among the legions of dispossessed had begun to grow in the mid-seventies, and since 1989 the state has responded with increasingly harsh repression. Overall, as many as 1.3 million people have been killed and 3 million displaced by war and war-related famine since 1983.[45]

Throughout the Sahel, pastoralists are under pressure—from state-sponsored mechanized farming and commercial ranching projects, governmental and commercial restrictions on access to traditional rangelands, and the effects of population growth. A 1991 study of the Horn of Africa by the World Conservation Union found that "the disenfranchisement of local peoples from traditional land and water rights has been a major factor fuelling conflict and instability." Unable to affect the larger forces at work, in several cases they have struck against closer targets—other groups of pastoralists or small farmers, all of whom find themselves in greater competition for land and water. Examples of this pattern can be found in Ethiopia, Kenya, Mali, Mauritania, Niger, and Senegal, as well as Sudan.[46]

* * * *

On the high seas, environmental depletion is leading to conflicts of a different nature and format. Given the severe mismatch between global fishing fleets and the ebbing numbers of cod, tuna, and other valued fish species, serious tensions over fishing rights have flared among a number of nations, and the possibility of governments resorting to force does not seem entirely farfetched. During the seventies, the United Kingdom, Iceland, and several countries engaged in what was dubbed the "cod wars," confrontations on the high seas between trawlers and gunboats.[47]

The 1982 Law of the Sea, which came into force in the mid-nineties, dealt with these issues by allowing coastal nations to regulate fishing inside exclusive economic zones (EEZ) stretching up to 200 nautical miles from their shorelines and encompassing the oceans' prime fishing grounds. Conflict now arises over migratory fish (such as tuna) that straddle EEZ lines. As the limits of the oceans' bounty are making themselves felt in the form of depleted stocks, coastal nations are pitched against those with substantial long-distance fishing fleets. Half the fish caught off West Africa are snatched up by sophisticated foreign fleets for which the local fishing vessels are no match.[48]

Taking place in remote waters, scuffles between fishing and military vessels of different nations usually evade the public eye. But when Canada seized the Spanish trawler

Estai outside its territorial waters on March 9, 1995, it triggered a crisis in relations between the two countries and made headlines around the world. That particular conflict arose over jockeying for national fishing quotas of the Greenland halibut (turbot). In 1994, the Northwest Atlantic Fisheries Organization had limited the total allowable catch and allocated national quotas as a conservation measure; when European governments rejected the agreement, however, Canada imposed a unilateral moratorium. The *Estai* seizure came as a warning to comply with the moratorium, but since it took place in international waters, it clearly was in violation of established law.[49]

Spain invoked the freedom of the seas principle; Canada presented itself as the agent of marine conservation. But both governments acted under strong pressure from their respective fishing regions and industries to protect what remained of jobs and income. In Canada's Newfoundland province, the unemployment rate shot up to 20 percent after years of overfishing caused the cod industry to collapse in the late eighties. As rhetoric grew more heated, Canada prepared to seize additional ships, and Spain sent military vessels to escort its boats. An escalation was narrowly averted, however, as Canada and the European Union reached an agreement that not only established new quotas, but also provided for enforcement and monitoring, both onboard and by satellite, to ensure that conservation rules (such as minimum sizes of fish to be caught) are complied with.[50]

The clash over the *Estai* is far from an isolated case, however, and the underlying issue—nations locked into a zero-sum game over an increasingly depleting resource—remains to be fully addressed. In September 1994 alone, reports came in that a French fisherman had been shot by a crew member of a Spanish boat, that the Icelandic coast guard had started escorting fishing vessels in the Barents Sea to protect them from impoundment by the Norwegian coast guard, and that Russia had jailed three Japanese fishers for illegal fishing.[51]

Ongoing conflicts over fishing rights in the open seas were of sufficient concern to some 100 countries to prompt them to open negotiations under the aegis of the United Nations in July 1993. Two years later, a treaty to deal with the depletion of straddling and migratory fish was approved (although it needs to be ratified by at least 30 nations before taking effect). It establishes a legally binding framework within which more specific rules for conservation are to be elaborated at regional and local levels. Fishing fleets will have to provide data on the size of their catches and take steps to reduce overfishing and minimize the currently huge proportion of the catch that is discarded as waste. There are also provisions for verification, dispute settlement, and enforcement.[52]

Environmentalists criticize the treaty as inadequate in some respects, but agree that several provisions "represent significant advances in international fisheries law," as a Greenpeace analysis stated. In particular, they endorse the "precautionary approach" that instructs signatories not to use the absence of conclusive scientific data as an excuse for failing to establish conservation measures. It remains to be seen how well this agreement will be implemented.[53]

* * * *

The degradation of land, freshwater, and marine resources will become more pronounced with the onset of global climate change, and environmentally induced conflicts are likely to intensify. But in addition to disputes in these individual areas of concern, the global politics of climate change itself is already becoming a focus of diplomatic and political conflict. Ever since the Rio conference in 1992, countries have been maneuvering to limit or avoid responsibility for reducing their emissions of carbon dioxide and other greenhouse gases.

To a large extent, this is shaping up as a dispute between the industrialized North, accounting for the bulk of emissions, and the South, which would primarily be a victim of global warming. Developing nations have been clamoring for technical and financial assistance to cope with the consequences. On the horizon is an issue that has barely been addressed yet: with developing countries, particularly China, planning to expand their use of fossil fuels, industrial countries will need to slash their emissions even more dramatically, or they will need to develop and share alternative energy technologies much more generously. But the politics of global warming is characterized by a sharp discrepancy between the status quo of sovereign nation-states and the growing reality of borderless and massive environmental change. (See Chapter 7.)

As this chapter has illustrated, environmental degradation or depletion usually is one of a series of stress factors that, in complex cause and effect, may combine to trigger violent confrontation or exacerbate already raging conflicts. Typically there is no such thing as an exclusively "environmental conflict." Even in Bougainville, other factors—economic and ethnic—played a role. Environmentally induced conflicts display an extraordinary variety of circumstances in terms of the nature of environmental change involved, the intervening social and economic factors, the range of protagonists, and the conduct and impact of the conflict. The social and economic stress factors referred to repeatedly thus far are considered further in Chapter 4.

4

Inequality and Insecurity

We have less insecurity in the military sphere and more insecurity in the personal and community spheres. We've replaced the threat of the nuclear bomb with the threat of a social bomb.[1]

Juan Somavia, Chile's U.N. Ambassador and Secretary-General, World Social Summit, Copenhagen, Denmark, March 1995

"How are we supposed to live on such low salaries?" asked Javier Gonzalez, one of tens of thousands of Bolivian teachers, health care workers, miners, and oil industry workers on strike in March 1996 in the nation's capital, La Paz. "We won't stand for it," he continued: "Fight, fight, fight!" Strikes in Bolivia by workers pushing for an increase in wages and opposing the government's economic policies have grown in size in recent years. In 1995, the government imposed a state of emergency in response, and detained union leaders. In Bolivia, as elsewhere in much of Latin America, many people are hard-pressed to find jobs or to make do with the often meager incomes that existing jobs give them. Low-paid jobs are now often filled by middle-class people laid off from better-paying jobs.[2]

Yet on the face of it, things look promising. Economic growth in Latin America was so strong in the early nineties that the economies of the region came close to making up for the fall in gross national product (GNP) during the "lost decade" of the eighties—years marked by the onset of a severe debt crisis and recession. As foreign capital pours into the region—a fourfold increase between 1990 and 1993—in response to privatization, trade liberalization, and deregulation, new markets are emerging and new opportunities beckon.[3]

The upturn in these macroeconomic indicators misleadingly suggests that the well-being of the population is improving across the board. But with few exceptions, income distribution in the region remains more skewed now than it was before the start of the debt crisis. The United Nations Economic Commission for Latin America and the Caribbean reckons that even though economic growth is expected to continue through the nineties, the region's poverty rate will not drop below its 1990 level of 46 percent; in fact, it may increase slightly. In Chile, often held up as a model of free market reforms, poverty rates are only now declining to a level last seen in 1970. The incomes of some 192 million Latin Americans are below the poverty line, and almost half these people are extremely poor; 130 million people are homeless or live in unfit housing structures. A front-page *New York Times* headline in late 1994 summarized the situation well: "Latin Economic Speedup Leaves Poor in the Dust."[4]

Though the circumstances vary from region to region, Latin America is not alone in its experience of a highly uneven distribution of the benefits of economic growth (or the woes of economic contraction). Inequality, marginalization, and the resulting polarization in society appear to be on the march virtually worldwide. Many societies, including those misleadingly called "developed," are confronting the paradox of a growing GNP and stagnating or eroding incomes and living standards.

The challenge is not just how to alleviate the poverty of the downtrodden. In some countries, even members of the middle class are becoming less secure. Technological development has increasingly allowed a hyper-productive economy able to churn out huge amounts of industrial and agricultural products with the help of fewer and fewer workers, giving rise to the notion of "jobless growth." Employment is becoming more precarious with declining job security and the increasing phenomenon of part-time and temporary work. Societies have been far less adept at spreading the benefits of economic growth than they have been at unleashing productive forces.

★ ★ ★ ★

Accelerating economic globalization primarily benefits countries, communities, companies, and individuals with capital, technological know-how, and entrepreneurial skills. And it primarily suits countries that are already well positioned in the global economy. The economies of East and Southeast Asia are poised to take advantage of the opportunities that further global economic integration promises, rapidly shedding the status of "developing" countries and joining ranks with the "old" industrial nations. Latin America, in a relatively promising position prior to the debt crisis, is still struggling to overcome the legacy of the eighties.

Much of sub-Saharan Africa, in contrast, has long been a marginal and extremely vulnerable player in the global economy. The region has attracted only a tiny share of foreign investment, and most of its countries are in danger of being increasingly bypassed. Of the global foreign direct investment flows of $224 billion in 1994, only $3.1 billion—1.4 percent—went to African countries. Under the latest trade liberalization measures, known as the Uruquay Round, African countries are projected to lose more than $2 billion in annual exports as their preferential access to European markets evaporates.[5]

A good deal of the global rich-poor gap is embodied in the persistent North-South disparity: the developing world accounts for three quarters of the world's population but has only 16 percent of global income. But gaps exist also within countries of both the South and the North. (See Table 4–1.) There is now a "Third World" even within the richest countries, just as there is a "First

TABLE 4–1. *Ratio of Richest 20 Percent of Population to Poorest 20 Percent in Selected Countries, 1981–92*

Country	Ratio
Brazil	32
Guatemala	30
Senegal	17
Mexico	14
United States	13[1]
Malaysia	12
Zambia	9
Algeria	7
China	7
South Korea	6
Germany	6
India	5
Japan	4

[1]Data for 1993.
SOURCE: See endnote 6.

World" within poorer ones. As a group, Latin American countries have long displayed the most unequal income distribution in the world—disparities that grew even bigger during the eighties. In Argentina, Brazil, Costa Rica, Uruquay, and Venezuela, for example, the richest 5 percent gained during the eighties at the expense of the bottom 75 percent.[6]

For people at the bottom of the global economic heap, particularly in developing countries, the day-to-day reality is less one of dazzling possibilities than of innumerable hardships and chronic insecurity. They contend with meager incomes despite long hours of backbreaking work, insufficient amounts of food and poor diets, lack of access to safe drinking water, susceptibility to preventable diseases, and housing that provides few comforts and scant shelter.

Rich-poor disparities are about much more than just the inability to accumulate material wealth: they are often a matter of life and death. The 1991 edition of the *Human Development Report* noted that the life expectancy of the poorest Mexicans—53 years—was 20 years less than that of the richest of their compatriots. And in rural Punjab, a region straddling India and Pakistan, babies born to landless families have infant mortality rates one third higher than those born to landowning families. Despite undeniable improvements in living standards and health and education since mid-century, massive numbers of people, mostly in developing countries, remain mired in poverty, with some of their most basic needs unmet. (See Table 4–2.)[7]

The growing gap between rich and poor generates discontent and resentment, and it fuels social conflict. As the *New York Times* asked rhetorically, reporting in late 1994 on the seemingly sanguine macroeconomic trends

TABLE 4–2. *Dimensions and Magnitude of Human Insecurity, Early Nineties*

Source of Insecurity	Observation
Income	1.3 billion people in developing countries live in poverty; 600 million are considered extremely poor; in industrial countries, 200 million people live below the poverty line.
Clean Water	1.3 billion people in developing countries lack access to safe water.
Literacy	900 million adults worldwide are illiterate.
Jobs	820 million people worldwide are unemployed or underemployed.
Food	800 million people in developing countries have inadequate food supplies; 500 million of them are chronically malnourished, and 175 million are children under the age of five.
Housing	500 million urban dwellers worldwide (out of 2.4 billion) are homeless or live in inadequate housing; 100 million young people are homeless ("street children").
Preventable Death	15–20 million people die each year due to starvation and disease aggravated by malnutrition; 10 million people die annually due to substandard housing, unsafe water, and poor sanitation in densely populated cities.

SOURCE: See endnote 7.

in Latin America: "But if things are so rosy, why did peasants rise up this year in southern Mexico? Why has Venezuela had two coup attempts and continued unrest? Why have Bolivian workers staged national strikes? And why, in Argentina, considered a stellar example of economic transformation, did workers burn a provincial government building last December and march on the capital this summer?"[8]

Orthodox economic theory has it that "a rising tide lifts all boats"—that is, economic growth improves the livelihoods of everyone. Yet the gap between rich and poor has grown to tremendous levels. The richest fifth of the world's population now receives more than 80 percent of global income, while the poorest fifth has to make do with a little more than 1 percent. In 1960, those in the top 20 percent had 30 times the income of those in the bottom 20 percent; by the beginning of the nineties, they had almost 60 times as much. The world's 358 billionaires had a combined wealth of $762 billion in 1994—the equivalent of the income of 2.4 billion people, 45 percent of the global population.[9]

Western industrial countries have vastly greater resources to "lift all boats," but even so some 100 million people there—more than 10 percent—live below the poverty line, and more than 5 million are homeless. About 15 percent of Americans—39 million people—were officially considered poor in 1993. And 100 million people in formerly Communist industrial countries live in poverty. Inequality has increased in many rich nations. In the United Kingdom, the ratio between the top 20 and bottom 20 percent went from 4:1 in 1977 to 7:1 in 1991. In the United States, which has the widest income gap among industrial nations, it went from 4:1 in 1970 to 13:1 in 1993.[10]

The nominal wealth of a country is not always a good indicator of its peoples' well-being, however. A low GNP does not have to mean a low score in terms of human development, as the *Human Development Report 1995* shows for Costa Rica, Madagascar, Vietnam, and several other countries, or as is true for the state of Kerala in India. And a high GNP does not necessarily mean that the social fabric is not subject to substantial wear and tear. An annual compilation of 16 indicators by the Fordham Institute for Innovation in Social Policy to measure social health in the United States suggests that social well-being has declined since the mid-seventies, even as GNP has followed a steady upward trajectory.[11]

* * * *

In most developing countries, where agriculture is a mainstay of the economy and key to people's livelihoods, land distribution is an indicator as important as the distribution of wealth. It, too, tells a story of immense imbalances. Large majorities of rural inhabitants are landless or have too little land to sustain themselves. Many are forced to migrate to more marginal areas, such as hillsides and rain forests, that are easily susceptible to erosion and whose soils are quickly exhausted. (See Chapter 2.) In Mexico, for example, more than half of all farmers are eking out a living on land on steep hill slopes that now account for one fifth of all Mexican cropland. Others turn to seasonal or permanent wage labor on large agricultural estates, or end up seeking new livelihoods in already crowded cities.[12]

On the whole, Latin American land tenure patterns are much more inequitable than those found in Asia and Africa. (See Table 4–3.) Unequal landownership is of course nothing new—in Latin America, it is an enduring legacy of colonialism—but in more recent years the mechanization of agriculture in some areas has led to the eviction of millions of small peasants and sharecroppers by commercial farmers—as is the case in Sudan, discussed in Chapter 3. The Institute for Development Studies in the United Kingdom estimates that 90 percent of the marketable agricultural production in Sudan is controlled by fewer than 1 percent of its farmers. To-

TABLE 4–3. *Land Distribution and Landlessness, Selected Countries or Regions*[1]

Country/ Region	Observation
Brazil	Top 5 percent of landowners control at least 70 percent of the arable land; the bottom 80 percent have only 13 percent of the cultivable area; 12 million rural Brazilians are landless or near-landless, yet enough land is currently left idle by large landowners to provide the 12 million with more than 2 hectares of land each.
Peru	Three quarters of the rural population is landless or near-landless.
Central America	In Guatemala, 2 percent of farmers control 80 percent of all arable land; in Honduras, the top 5 percent occupy 60 percent; in El Salvador, the top 2 percent own 60 percent, and almost two thirds of the farmers are landless or nearly landless; in Costa Rica, the top 3 percent have 54 percent of the arable land.
India	40 percent of rural households are landless or near-landless; the 25 million landless households in 1980 are expected to reach 44 million by the end of the century.
Philippines	3 percent of landowners control one quarter of the land; 60 percent of rural families have no or too little land.

[1]Near-landlessness means that a rural family or household possesses too little land to sustain its members' livelihoods with farming alone.
SOURCE: See endnote 13.

gether with population growth, which forces peasants to subdivide plots into smaller and smaller parcels from one generation to the next, unequal land tenure is causing increasing landlessness. In 1981, an estimated 167 million households (containing 938 million people) were landless or near-landless, and their numbers are expected to increase to nearly 220 million by the turn of the century.[13]

★ ★ ★ ★

For many developing countries in Latin America and Africa, the sharp increases in what were already large social and economic discrepancies were a consequence of structural adjustment programs imposed by the International Monetary Fund and the World Bank since the early eighties. There is little doubt that many of these economies needed adjustment (although the type of adjustment most suitable is very much in contention): they suffered from a combination of home-grown and external problems, including distorted economic structures, wrongheaded economic policies, careless borrowing of foreign funds, soaring interest rates in the late seventies, and plunging world market prices for their commodity exports in the eighties.

The drop in world prices led to an erosion of the terms of trade—the ratio of export prices measured against import prices on a unit-per-unit basis. In several countries, warfare worsened problems by absorbing large amounts of money and disrupting local economies. The upshot of all these factors was that foreign debt soared from $217 billion in 1970 to $1.1 trillion in 1980 (expressed in 1993 dollars)—the equivalent of 88 percent of the Third World's annual exports, or 27 percent of these counties' GNP.[14]

By the mid-nineties, some 75 countries had signed on to structural adjustment programs. These typically require recipients of adjustment loans to implement measures such as lowering trade and investment barriers; devaluing the currency; reducing or eliminating subsidies, price controls, and social programs; and privatizing state enterprises. This proved to be exceedingly bitter medicine for the poor and even for large parts of the middle class. And yet the severe shock therapy failed to do what in theory it should have: put in place more successful and sustainable economic structures. The U.N. Research Institute for Social Development concluded in 1995 that most structural adjustment experiences had been failures, and that even those that yielded some of the expected macroeconomic results had detrimental social consequences.[15]

One reason for the failure of adjustment was itself structural: it initiated a vicious cycle of retrenchment and disinvestment that led to economic stagnation or decline, rather than putting in place the building blocks for a new beginning. In sub-Saharan Africa, for instance, investment declined by 50 percent during the eighties. Nor did structural adjustment loans lighten the debt burden; intended to help debtor countries to service their debts (and hence protect lenders from the perils of default), they instead kept them on the debt treadmill. Total Third World foreign debt climbed from $1.1 trillion in 1980 to close to $2 trillion in 1994 (in 1993 dollars). Walden Bello, executive director of the Institute for Food and Development Policy in Oakland, California, notes that "between 1982 and 1986, Third World countries received $25 billion more from official creditors than they paid out to them, while they paid the commercial banks $183 billion more in interest and amortization than they received in new bank loans." The U.S.-based Bread for the World notes that African governments spend more than twice as much servicing their debts as they do on health and primary education combined.[16]

Most people in highly indebted African and Latin American countries suffered a severe drop in living standards during the eighties—hence the term "lost decade."

Latin American per capita income fell by 11 percent. And in sub-Saharan Africa, per capita income by the end of the eighties was no higher than at the time of independence in the early sixties.[17]

In Mexico, real wages declined by more than 40 percent in 1982–88. In 1983, a basket of basic goods for an average family of five cost 46 percent of the minimum wage; by 1992, it cost the equivalent of 161 percent. In Nicaragua, declining incomes led to a 15-percent drop in per capita consumption of such basic staples as corn and beans between 1990 and 1993. Observers might simply chalk this up to the difficulties associated with an economy still reeling from years of warfare. But even neighboring Costa Rica, widely regarded as an island of stability in a troubled region, has suffered: in a sharp reversal of its historical record, real wages there declined 17 percent between 1980 and 1991, and poverty increased from 21 to 28 percent in 1987–91.[18]

Not surprisingly, the hardships and deepening disparities have triggered or reopened political rifts within these societies and provoked a wave of urban protests. In Mexico, impoverished peasants in the southern state of Chiapas rose up in revolt in early 1994 (see Chapter 6), while increasing numbers of residents choose illegal migration to the United States as their ticket to a better life (see Chapter 5). In Nicaragua, desperate workers have unleashed a wave of strikes since 1990. Many ex-Contra guerrillas and former Sandinista government soldiers have taken up arms again—sometimes joining forces—to press their demand that the government honor social and economic commitments made under the country's peace agreement. And in Costa Rica, increasing economic polarization may begin to jeopardize the country's stability and democratic tradition.[19]

Structural adjustment policies are part of a larger, neoliberal model of development that sees privatization, deregulation, trade liberalization, and world market integration as the road to economic salvation. In agriculture, this has meant giving priority to the larger cash crop producers (who are typically more oriented toward lucrative export markets and nonstaple and perhaps even nonfood crops) to the detriment of the numerically much larger subsistence or small-scale commercial farmers (who are often seen by technocratic leaders as inefficient and an obstacle to modernization). In many countries, small farmers are losing access to credits, extension services, and other forms of support, such as guaranteed prices. The 1994 edition of the *Human Development Report* reported that in many developing countries, 40 percent of the people typically receive less than 1 percent of the total credit disbursed.[20]

There is also evidence that large, often export-oriented, commercial operators rather than small farmers are being given priority access to fertile land and water. In the Jodhpur district of Rajasthan, India, increasing use of scarce groundwater to cultivate chili peppers and other water-intensive cash crops has caused village wells used by the rural poor to go dry. Villagers saw their land degrade, their herds die, and their communities fall apart. Similarly, in Colombia, flower production for foreign markets has caused groundwater levels to fall, harming local food production. In Senegal and Mali, fruit and vegetable export plantations were developed to the detriment of the peasant economy.[21]

International markets no doubt hold considerable opportunities for some wealthier farmers, but many small commercial farmers and particularly subsistence peasants are struggling to survive; they often lack access to credit and do not have the resources required to modernize and intensify their operations in order to compete in the brave new world of globalized agriculture. With import tariffs being lowered now that agriculture is to be opened up more to international trade, these producers increasingly compete with a flood of cheap grain imports. Some 5 million poor Brazilian peasant households, for example, see their very existence threatened in this manner: Brazil's wheat imports surged sixfold between 1988 and 1995; they now supply 79 percent of the country's consumption.[22]

The implications for social stability are stark. The frictions between subsistence or near-subsistence peasants and commercial farms may lead to intensified social conflict in the countryside and perhaps to uprisings, as happened in Chiapas. Or marginalized peasants, already facing environmental and demographic pressures, may join the trek to urban areas, where they add to the strain on infrastructure, social services, and jobs.

★ ★ ★ ★

It would be one thing if those displaced in the countryside could find jobs in cities and towns. But by and large, these jobs do not exist. Indeed, one reason for rising inequality and poverty—and a major threat to social cohesion and stability—is found in what various observers have termed the global jobs crisis. Out of the global labor force of about 2.8 billion people, at least 120 million people are unemployed, while 700 million are classified as "underemployed"—a misleading term because many in this category are actually working long hours but receiving too little in return to cover even the most basic of needs. The record shows that economic growth per se does not ensure that everybody who is willing to work will find a job: between 1975 and 1990, economic output worldwide grew 56 percent, but employment only by half as much.[23]

Unemployment, underemployment, the threat of job loss, and the specter of eroding real wages are challenges for many workers across the globe, though the particular conditions and circumstances diverge widely in rich and poor countries. In advanced industrial countries, three phenomena can be observed. First, technological development, particularly the rise of microelectronics, has dramatically reduced the need for labor—so much so that a substantial share of unemployment in these countries now is structural rather than shaped purely by the growth-recession cycle.

Second, many companies are embracing measures—such as smaller and more flexible production units, subcontracting work, and temporary or part-time hiring—that allow them to adapt rapidly to fast-changing market conditions but render job tenure more tenuous and insecure. Third, corporations are increasingly able to tap into a large pool of cheap labor in developing countries, replacing a much higher paid domestic work force—due to modern communications and transportation networks, the ability to parcel out components of the work process, and increased capital mobility. Initially, unskilled or semiskilled jobs were at risk in this manner, but recent evidence suggests that skilled workers are now facing similar pressures.[24]

A key effect of this cheap labor strategy is strong downward pressure on wages, as evidence from the United States makes clear. Between 1973 and 1990, real wages for production or nonsupervisory workers (excluding agriculture) declined by more than 20 percent. Although the loss of 43 million jobs between 1979 and 1995 was numerically more than offset by some 70 million newly created jobs, two thirds of those finding new employment now earn less. In fact, an estimated 18 percent of full-time U.S. workers now have jobs that pay wages at or below the poverty level.[25]

In contrast to the United States, many other industrial countries have, so far at least, not embraced the low-wage strategy—for fear of rapidly growing economic inequality among their populations and the implied threats to the social and political health of their societies. But they face high and growing unemployment rates that not only burden the welfare state but gnaw at the foundations of social stability. Unemployment stands at 10 percent in the United Kingdom, 12 percent in France and Italy, and 22 percent in Spain. At 11 percent, German unemployment reached a postwar record high in early 1996. (U.S. unemployment stood at 6 percent in 1994, though some analysts say revisions in the official count left out many people who should be included.) Among members of the Organisation for Economic Co-operation and Development (OECD), Japan alone has managed to keep joblessness low—at slightly above 3 percent; this is nevertheless the highest it has been since the end of World War II.[26]

The social and psychological impacts are often traumatic. People unable to find work, or work only at substantially lower wages than they are accustomed to, suddenly see their comfortable middle-class existence evaporate: their careers are interrupted or brought to a premature end, their incomes are no longer sufficient to meet the bills, and their retirement security is beset by uncertainty. In modern societies, a job not only provides the means to pay the bills, it often helps people define who they are. Joblessness carries a heavy stigma in a success-oriented, materialistic society. While some cope well, others become depressed, develop illnesses, or abuse alcohol or drugs; frequently, their relationships with spouses, children, and friends suffer. The threat to jobs is such that even those who survive one round of layoffs feel insecure, wondering if they are next.[27]

It is difficult to predict the full consequences of the current jobs crisis. But it seems likely that the society-wide effects will be greater than the sum of its individual parts. Whether the current situation is compared with the Industrial Revolution or perhaps the tumultuous Depression Era, the implication is clear: failure to deal appropriately with sharpening social problems could have fatal political consequences. People whose hopes have worn thin, whose discontent is rising, and whose feelings of security have been stripped away are more likely to support extreme "solutions," and it is clear that some politicians stand ready to exploit the politics of fear. Alarmism is uncalled for, but it is prudent to remember that the hardships and dislocations of the Great Depression resulted in the rise of fascism and a world war.[28]

Even though workers in OECD countries are clearly the global "elite" in terms of wages and benefits, the threat to their job and wage security has already triggered two reactions: calls for protectionist policies against imports from countries where labor is cheap, and hostility toward immigrants seen as taking jobs or social benefits away from domestic workers. It matters less whether these perceptions are correct than that they are clearly helping to fan antiforeigner sentiments and hatreds that have led to violence and that can generate explosive social and political conditions.[29]

Whereas western countries have experienced a gradual rise in unemployment during the past two decades, the formerly Communist countries of Eastern Europe and the Soviet Union have had to contend with rapidly emerging mass unemployment and dramatic increases in poverty and inequality. They continue to undergo a wrenching and uncertain transition to what must seem like a highly uncertain future from a system that, although highly inefficient and even demoralizing, provided a sheltered kind of employment and a sense of steadiness. Suddenly being forced to compete on world markets with more efficient western companies, many enterprises have gone bankrupt. From 1990 to early 1992 alone, the ranks of the unemployed in these countries rose from 100,000 to more than 4 million. From virtually zero in the late eighties, by 1992/93 unemployment rates had climbed to 14 percent in Poland, 11 percent in Hungary, and about 5 percent in Russia. Real wages declined some 30 percent in Poland between 1989 and 1993 and 27 percent in Russia in 1991–93.[30]

The experience in different parts of the developing world has been sharply divergent. The dynamic economies of East and Southeast Asia have experienced high growth in productivity, output, employment, and real wages. Unemployment rates ranged between 2 and 3 percent during the late eighties and early nineties. Employment and wage levels in most countries in Latin America and Africa, by contrast, have suffered from structural adjustment, as discussed earlier.[31]

Many of those unable to find regular jobs drift to the informal sector—the underbelly of the economy of many developing countries. But this area is characterized by low skills, productivity, and pay (though some talented entrepreneurs can do well), and offers no form of social protection. In Africa, the International Labour Organisation (ILO) finds that "the majority of workers in the informal sector would be very fortunate to earn even the official minimum wage." According to ILO, in sub-Saharan Africa the informal sector employed more than 60 percent of the urban work force in 1990. Its share of the nonagricultural work force in Latin America rose from 40 to 53 percent during the eighties.[32]

Perhaps most unsettling is the reality of large-scale youth unemployment, which virtually everywhere is substantially higher than that for the labor force as a whole. One survey of 15 African countries showed youth unemployment rates to be triple those for adults. Even in most industrial countries, youth unemployment is an enormous challenge: in the early nineties, it reached 14 percent in the United States, 15 percent in the United Kingdom, 26 percent in Italy, and 36 percent in Spain. (Japan and Germany are the exceptions, with rates of 5 and 6 percent, respectively.)[33]

Yet high rates of population growth and the resulting disproportionately large share of young people in many developing countries translate into much greater pressure on job markets there. Roughly 20 percent of the population in industrial countries is age 15 or younger. But in China, the figure is 27 percent; in Latin America, 34 percent; in South and Southeast Asia, 38 percent; and in Africa, 45 percent. The uncertain prospects that many young adults face are likely to provoke a range of undesirable reactions: they may trigger self-doubt and apathy, cause criminal or deviant behavior, feed discontent that may burst open in street riots, or foment political extremism.[34]

The world's labor force is projected to grow by almost 1 billion during the next two decades, mostly in developing countries hard-pressed to generate anywhere near adequate numbers of jobs. During the nineties, an additional 38 million people will seek employment each year in these countries.[35]

* * * *

Inequality, poverty, and lack of opportunity are, of course, nothing new. But today's polarization takes place when traditional support systems are weakening or falling by the wayside. In developing countries, there is an erosion of the bedrock of social stability—the webs and networks of support found in extended family and community relationships (although these are admittedly often paternalistic and exploitative). It is unclear what will take its place.

In western industrial countries, the post–World War II welfare state substituted or supplemented many family and community support functions, providing an extraordinary degree of social stability. Since the seventies, however, the welfare state has come under growing attack, both because of ideological objections and because of stagnating public revenues in the face of continuously growing welfare costs. Among former Eastern bloc states, the political upheaval of the late eighties succeeded in loosening and then demolishing the grip of authoritarian institutions, but creating viable alternatives has proved far more elusive.[36]

Privatization and economic globalization have unleashed entrepreneurial energies and led to the emergence of vast new markets and opportunities. Yet this type of progress has also brought profound disquiet. Uncertainty about what the future holds and fears of losing a job or livelihood are now endemic. No country, company, community, or individual can rest on past success, and many are struggling hard just to avoid losing ground. "Global markets [are permitted to] wreak havoc with the livelihoods of many of the world's people," in the words of a 1995 U.N. report entitled *States of Disarray*. The social fabric of many societies, whether affluent or destitute, is under greater strain as the paths of rich and poor diverge more sharply, and as the stakes of economic success or failure rise.[37]

5

People on the Move

Trying to escape the hardships imposed by cyclones, flooding, drought, and land scarcity, large numbers of Bangladeshis have been migrating during the past four decades to Assam, Tripura, and other northeastern states in neighboring India. Counting the people who have followed this route is highly controversial: officially, the Bangladeshi government maintains that there has been no large-scale migration, but others estimate that anywhere from 12 million to 17 million people—roughly 10 percent of the total population—have moved illegally since 1951. Bangladeshis are attracted to northeastern India not only because of its proximity, comparable climate, and fertile soils, but also because the region is far less densely populated: 284 persons per square kilometer in Assam and 262 in Tripura at the beginning of the nineties, compared with more than 900 in Bangladesh.[1]

Indian politicians from several parties facilitated the illegal immigration because it allowed them to build up clientele "voting banks." Although the migrants were accepted initially by the Assamese, they were increasingly seen as a threat. Concerns about religious and cultural identity played a role, but economic and political factors were no less important. Social tensions rose as the migrants acquired more land and began to prosper.[2]

In 1979, an anti-immigrant movement emerged in Assam. Peaceful protests gave way to violence. In 1983, perhaps as many as 5,000 Bangladeshi immigrants were killed in a series of massacres. An anti-immigrant party, Assam Gana Parishad, managed to unseat the ruling Congress party in state elections in 1985, though it subsequently failed to deliver on its promise of deporting immigrants.[3]

The phenomenon of local people having a change of heart regarding newcomers has not been confined to Assam. Migrants and refugees in many areas are finding that they are unwelcome in their "host" countries: In Germany, right-wing groups have engaged in repeated acts of violence or arson against Turkish workers and foreigners seeking political asylum. In the United States, the videotaped beating of several Mexicans by police officers in early 1996 was symptomatic of rising anti-immigrant sentiments. It set off a storm of protest in Mexico, yet that country deported illegal settlers from its own territory in 1995. In the Russian Far East, the government has launched several sweeps against Chinese migrants. In Africa, Gabon forced more than 50,000 migrant workers to leave the country in early 1996, and Libya deported large numbers of African migrant workers, accusing them of "spreading AIDS and stealing jobs."[4]

Anti-immigrant feelings reflect the many pressures and insecurities that people everywhere are experiencing. As noted in earlier chapters, ours is an age in which global competition has vastly increased people's sense of economic vulnerability, in which social services and benefits are being pruned or slashed, and in which scarce land and water are increasingly being contested. Under such circumstances, the influx of large numbers of people is all too easily seen as menacing. Migrants or refugees are perceived as taking away jobs, imposing economic burdens, irredeemably altering the local culture and customs, or generally becoming unwelcome competitors for scarce resources and services. The experience of recent years shows how much political leaders in many countries have been tempted to use the migration issue to seize or maintain political power.

It would be misleading to suggest an inevitable linkage between population movements and violent conflict. Huge numbers of people have been absorbed into new host countries—some accepted temporarily, others integrated for good. But there are many cases in which resentments breed to the point of precipitating violence. Unless some of the underlying trends and conditions are checked, there may well be many more violent incidents in years to come.

★ ★ ★ ★

An unprecedented number of people are now moving both within countries and across borders. Sometimes they move in response to real or perceived opportunities in other countries, and by doing so they may benefit both their home and their host. In the host countries, migrant workers often become part of the "underbelly" of the economy—fulfilling important, though frequently scorned, menial and low-skilled tasks. And by sending remittances back home, these workers make important contributions in several countries. (See Table 5–1.)[5]

But population movements are as likely, if not more likely, to be a symptom of unresolved problems as they are an attempt to grasp new opportunities. To a very large extent, they are a consequence of the accelerating social, economic, and environmental pressures assessed in earlier chapters. The magnitude and speed with which movements of people now take place makes them important potential contributors to conflict and hence a factor in the post–cold war security equation.

TABLE 5–1. *Remittances by International Migrant Workers, Selected Countries, Early Nineties*[1]

Country of Migrants' Origin	Net Workers' Remittances	Remittances as a Share of Foreign-Exchange Earnings
	(million dollars)	(percent)
Albania	278	57
El Salvador	789	37
Egypt	4,960	31
Jordan	1,040	26
Bangladesh	942	26
Sudan	n.a.	25
Morocco	1,945	23
Mali	87	23
Greece	2,360	16
Portugal	3,844	15
Pakistan	1,562	15
Philippines	279	13
Turkey	2,919	10
India	3,050	9

[1]Monetary transfers by migrant workers to their home countries.
SOURCE: See endnote 5.

The number of international refugees rose from slightly more than 1 million in the early sixties to about 3 million in the mid-seventies, and then soared to an estimated 27.4 million in 1995. But at least another 27 million people worldwide—and possibly many more—have been displaced within their own countries. These numbers are still conservative; they do not include people uprooted by environmental calamities or "oustees"—those displaced by large-scale infrastructure projects. Over the past decade, for example, as many as 90 million people may have lost their homes to make way for dams, roads, and other "development" projects. Official definitions of what constitutes a refugee and who therefore is eligible for assistance and protection are outdated and overly narrow.[6]

A relatively small number of countries constitute the major sources of refugee flows (see Table 5–2): the top 10 account for 40 percent of all international refugees worldwide. The same ratio holds for the refugee-hosting countries, even though some 70 countries now have at least 10,000 refugees in their territories. Among the top 10 countries of asylum, only two are rich, industrial nations. Many of the poorer host countries have few resources at their disposal to assist refugees and may themselves suffer from symptoms of stress that could turn them, too, into refugee-generators.[7]

Indeed, several countries are in the peculiar situation of simultaneously being a source of refugee outflows and a host to displaced persons from neighboring countries. Sudan, for example, hosts people who have fled their homes in Chad, Eritrea, Ethiopia, and other countries, while Sudanese have sought refuge in Ethiopia, Kenya, Uganda, Zaire, and elsewhere.[8]

That refugees account for several percentage points of the population of even a large country like Iran, with

TABLE 5–2. *Largest Refugee-Generating Countries and Countries Hosting the Largest Refugee Populations, January 1995*

Refugee-Generating Countries			Refugee-Hosting Countries		
Country	Refugees		Country	Refugees	
	(thousand)	(percent of population)		(thousand)	(percent of population)
Afghanistan	2,744	15	Iran	2,236	4
Rwanda	2,257	29	Zaire	1,724	4
Liberia	794	26	Pakistan	1,055	1
Iraq	702	3	Germany	1,005	1
Somalia	536	6	Tanzania	883	3
Eritrea	422	12	Sudan	727	3
Sudan	399	1	United States	592	0.2
Burundi	389	6	Guinea	553	9
Bosnia	321	9	Côte d'Ivoire	360	3
Vietnam	307	0.4	Ethiopia	348	1

SOURCE: See endnote 7.

61 million inhabitants, underscores the magnitude of international refugee flows. But the real impact on receiving areas is more pronounced than these national averages suggest. Refugees are far from evenly distributed throughout their country of asylum. In a handful of places, they outnumber their hosts: in Tanzania's Ngara district, for instance, there are more than two Rwandan refugees for each Tanzanian.[9]

People displaced within the borders of their own native countries typically receive much less attention or assistance than international refugees, even though they face comparable or worse hardships. Table 5–3 lists the 10 countries with the largest number of internally displaced people.[10]

Adding together the ranks of the internationally and the internally displaced, the total number of the uprooted

TABLE 5–3. *Countries with the Largest Number of Internally Displaced People, January 1995*

Country	Number[1]	Share of Total Population
	(million)	(percent)
Sudan	4.0	14
South Africa	4.0	9
Angola	2.0	17
Turkey	2.0	3
Bosnia	1.3	37
Rwanda	1.2	15
Liberia	1.1	37
Afghanistan	1.0	5
Iraq	1.0	5
Sierra Leone	0.7	16

[1]Due to poor information, these are rough estimates only.
SOURCE: See endnote 10.

amounts to an astounding 63 percent of the population of Liberia, about 45 percent of Rwanda and Bosnia, and 20 percent of Afghanistan. With such enormous dislocations, there is little coherent society left to speak of. Not only are entire areas depopulated while others become overcrowded, but economies are typically at a standstill and governance structures have collapsed. And the instability of countries so thoroughly shorn of their social fabric has severe, destabilizing repercussions for their neighbors. The conflict in Liberia, for example, has spilled over into Sierra Leone, which during the nineties has been plagued by a steady rise in violence and banditry that are only partly of domestic origin.[11]

The other source of instability from people on the move is migration. The number of cross-border legal migrants is estimated to have reached about 100 million

TABLE 5–4. *Countries with Largest Legal Migrant Populations, 1990*

Country	Migrant Population	Share of Total Population
	(million)	(percent)
United States	19.6	8
India	8.7	1
Pakistan	7.3	6
France	5.9	10
Germany	5.0	6
Canada	4.3	16
Saudi Arabia	4.0	26
Australia	3.9	23
United Kingdom	3.7	7
Iran	3.6	6
Côte d'Ivoire	3.4	29

SOURCE: See endnote 12.

worldwide, while illegal migrants number anywhere from another 10 million to 30 million. Table 5–4 shows the countries with the largest migrant populations. More than 100 countries are now experiencing major migration outflows or inflows, according to the International Labour Organisation. A quarter of these nations are simultaneously a source and recipient of migrants. Within countries, too, substantial flows of people can be seen, typically from rural to urban areas, and from poorer to more prosperous provinces. An estimated 20–30 million people migrate to cities within their own country each year.[12]

Traditionally, a sharp distinction has been made between migrants and refugees. Migrants are thought to leave largely of their own choosing, "pulled" by the prospect of a better job or higher earnings, whereas refugees are compelled to vacate their homes, "pushed" out by war, repression, or other factors beyond their control. Because they move involuntarily and without preparation, refugees are more likely to be a net burden, whereas migrants are seen as competitors for jobs.[13]

But the categories are becoming blurred. People are increasingly leaving their homes for a mixture of reasons—involving both fears and hopes, both voluntary and involuntary influences. As Hal Kane points out in a Worldwatch Paper, in some situations, migrants could be characterized as individuals who had the foresight to leave early, before local conditions deteriorated to the point where they were compelled to move—that is, before human rights violations become massive, before economic conditions turned wretched, and before environmental deterioration made eking out an existence impossibly burdensome.[14]

★ ★ ★ ★

A variety of forces are at work in uprooting people. They range from chronic violence and persecution to pervasive unemployment and disparities in wealth and income, to land scarcities and environmental degradation. Often, they operate not in isolation but in combination—a "cocktail of insecurities," in Kane's words. Among these factors, environment-related causes have to date received neither official recognition nor sufficient attention from the world's governments.[15]

People are turned into environmental refugees either by sudden "unnatural" disasters, such as the 1986 Chernobyl nuclear accident, that drastically reduce the habitability of an area or by the more gradual worsening of environmental conditions, which steadily undermines people's ability to grow sufficient food or maintain their health. Land degradation, water scarcity, and the threat of famine are powerful factors forcing people to move. The mid-eighties drought in the Sahel region, for instance, drove more than 2 million people out of Burkina Faso, Chad, Mali, Mauritania, and Niger. Desertification has uprooted one sixth of the populations of Mali and Burkina Faso. Many of these individuals ended up in cities and towns: during the seventies and eighties, the Sahel's urban population quadrupled. Of those who crossed state boundaries, many went to Côte d'Ivoire, one of the relatively more stable countries in the region.[16]

In Ethiopia, masssive deforestation and soil erosion combined with population growth, inequitable land tenure systems, and inefficient agricultural practices to force large numbers of peasants out of the highland farming areas. Since 1900, the forested area of the Ethiopian highlands has declined from about 40 percent to slightly over 4 percent, exposing soil to the forces of erosion. According to environmental consultant Norman Myers, the highlands have been losing topsoil at an annual rate of more than 1 billion tons (about seven times the rate in the United States, relative to the cropland base). Some of the migrants moved toward the Ogaden in southeastern Ethiopia. Their influx contributed to rising tensions over grazing and land rights in this contested area, which in turn became a factor in triggering the 1977 war between Ethiopia and Somalia.[17]

Extremely unjust patterns of land distribution in many countries (see Chapter 4), coupled with population growth, are forcing huge numbers of rural people to move to cities, onto marginal land, or into neighboring countries. As in the Horn of Africa, the movement of Salvadoran peasants into neighboring Honduras caused tensions and eventually, in 1969, helped trigger what became known as the "soccer war" (so named because the fighting started after a contentious soccer game between the two national teams).[18]

But conflict due to migration can also occur within countries. Vaclav Smil of the University of Manitoba predicts that huge numbers of people in several of China's interior provinces will migrate over the next two decades. The reason can be found in the large and growing disparities of both environmental quality and economic development between China's poverty-stricken hinterland and its booming coastal provinces. The northern interior is characterized by arid lands, highly variable rainfall, and soil easily susceptible to erosion. These adverse natural conditions have been magnified by environmental mismanagement.[19]

Smil points out that 10 northern interior provinces, home to about 40 percent of China's population, account for almost 80 percent of the country's soil erosion and two thirds of its severe water shortages. He expects that 20–30 million Chinese peasants will be displaced by environmental degradation during the nineties, and that at least another 30–40 million will be uprooted by 2025—a figure that could be much higher if climate change becomes a full-blown reality.[20]

Where would these people go? Most likely to the southern and coastal provinces, putting immense pressure on local governments. Already in the past several years, the coastal cities have been swamped with what Smil terms "unmanageable waves of unskilled peasant migrants seeking better economic opportunities." China now has a "floating" population of job seekers estimated at more than 100 million. Additional large-scale movements of people would likely heighten regional animosities. But attempts by the central government to block this migra-

could well provoke rural revolts, argues Jack Goldstone, a fellow at the Center for Advanced Study at Stanford University. These challenges occur in a context of enormous uncertainty regarding China's future: the twin policies of economic liberalization and political authoritarianism are creating manifold tensions and contradictions that seem to exacerbate rather than help resolve China's problems.[21]

Nobody knows how many people are being uprooted for environmental reasons, and there is not even an agreed-upon definition of environmental refugees. In 1988, Worldwatch researcher Jodi Jacobson conservatively put the number as at least 10 million, but given Vaclav Smil's estimates, it must now be considerably higher. If climate change becomes a reality, the ranks of environmental refugees will be far larger. Two thirds of the people in the world live within 60 kilometers of a coast, a portion expected to reach 75 percent by 2010. People living in river deltas, in coastal areas, and on small islands will be particularly vulnerable to sea level rise and to more frequent and destructive storms that will either render some areas far less habitable or simply submerge them.[22]

Bangladesh and Egypt are among the most vulnerable nations. Studies by the Woods Hole Oceanographic Institution in Massachusetts suggest that by 2050, about 17 percent of Bangladesh's land could be lost, displacing a comparable portion of the country's population. And by 2100, some 35 percent of the population there could be affected. Much depends, of course, on the actual extent of sea level rise, the rate of land subsidence in river deltas, adaptation measures, and other factors.[23]

Some 46 million people in coastal areas, mostly in developing countries, are currently at risk each year of flooding due to storm surges. According to the Intergovernmental Panel on Climate Change, a 50-centimeter sea level rise would increase this number to about 92 million, and a 1-meter rise would bring it to 118 million. This assumes no measures to adapt to climate change, but it also does not account for continued population growth, which will substantially increase the size of the at-risk population.[24]

Additional displacements in a "greenhouse world" will come from the detrimental effects that climate change is likely to have on agriculture. Although global agricultural production may not decline—positive and negative consequences of global warming may balance each other out—some regions could face serious problems. Vulnerabilities are strong in semiarid or arid parts of such regions as sub-Saharan Africa, northeastern Brazil, and South and Southeast Asia—where people often depend on already marginal or heavily used lands, and have few economic opportunities outside of farming. Although any estimates remain highly speculative at this point, Norman Myers believes that eventually as many as 50 million people may be uprooted by the unfavorable consequences of global warming.[25]

* * * *

Migrants (and to some extent refugees as well) can and do contribute to the well-being of their host countries—by making available their labor, introducing new skills, or enriching the local culture in culinary and other ways. But the influx of large numbers of people into another country can be a substantial burden—increasing competition over jobs, communal facilities, social services, land, and water—especially if it occurs over a short period of time or if the host country's economy is stagnant or in decline. Also, by altering the local ethnic or religious balance, migration has at times been a source of tension and instability—and occasionally violence.

A less recognized impact is that on the environment, and the consequences that has for relations between host and refugee populations. Requirements for wood—for cooking, heating, and construction of shelters—have led to massive deforestation in areas surrounding refugee camps, for instance. Many African countries already face acute shortages of fuelwood; the added demand by refugees worsens the situation. An extrapolation based on estimates by the Office of the U.N. High Commissioner for Refugees (UNHCR) suggests that international refugees and internally displaced people together consume some 25–40 million tons of fuelwood annually, leading to the deforestation of 70,000–100,000 hectares each year.[26]

A recent U.N. report notes that Rwandan refugees have caused "nearly irreparable environmental degradation" in parts of Tanzania and Zaire by cutting down large numbers of trees for fuelwood. Within nine months of their arrival, refugees at the largest Tanzanian camp had to walk 12 kilometers to reach the nearest source of fuelwood. In Zaire, Virunga National Park—declared a World Heritage site due to its wealth of rare species—has been particularly affected. In addition, the refugees' livestock are compounding an overgrazing problem. Crop yields are declining along with soil fertility in the receiving areas.[27]

So serious is the refugees' impact in Tanzania and Zaire that the United Nations asked donor countries in early 1996 to provide $70 million for a program to stop environmental damage, to assist reforestation, and to rehabilitate roads, health and education services, and other overtaxed facilities. Similar impacts are being felt elsewhere around the world wherever large numbers of refugees congregate.[28]

In areas already marked by scarcity, the presence of many refugees can sharply increase competition over resources, and lead to tense relations between refugees and their hosts. For example, intense competition for fuelwood has led to antagonism between Zairians and Rwandese refugees and between Bangladeshis and refugees from Myanmar (formerly Burma). "Disturbingly," UNHCR comments, "a growing number of low-income countries are now citing such problems as a justification for the exclusion or repatriation of refugee populations."[29]

Population movements have many impacts on receiving areas. But tension and violence may be caused less by these impacts per se than by political leaders or would-be leaders who exploit grievances arising from these impacts—be it by whipping up hatreds and prejudice against foreigners or by opportunistically organizing migrants into political vote banks.

Among western countries, a sense of the "West against the rest" seems to have infected a substantial portion of political leaders and the public alike. As one scholar, Myron Weiner, put it, "Citizens have become fearful that they are now being invaded, not by armies and tanks, but by migrants who speak other languages, worship other gods and belong to other cultures, and who, they fear, will take their jobs, occupy their land, live off the welfare system and threaten their way of life." The desire to shut out any unwanted elements reveals a fortress mentality hypocritically at odds with western pressure on developing countries to allow the free flow of goods, services, technologies, and money.[30]

★ ★ ★ ★

Recent years have seen growing efforts to restrict the flow of both migrants and refugees. From the end of World War II to the mid-seventies, industrial countries actively recruited foreign "guest-workers." But with the end of the postwar boom, the emergence of labor-saving technologies, and the rise in unemployment, most of these countries stopped this practice, adopted more restrictive immigration laws, and in some cases provided incentives for guest-workers to return home. These changes, however, came precisely when economic stagnation, population growth, rising unemployment, and social and political instability heightened the migration pressures in developing countries.[31]

Just as barriers against foreigners seeking jobs have arisen, so have restrictions against those seeking political asylum. Paul Macek, a Winston Foundation for World Peace Fellow, observes that "in contrast to the past, when refugees were welcomed in the ideological struggle between East and West, the uprooted increasingly face a hostile reception in host countries and are often turned away at the door."[32]

Some poorer countries have tried to close their borders. In the industrial world, meanwhile, UNHCR points out, "governments have tended to achieve similar results by more sophisticated means, either by interdicting asylum seekers...or by extending their immigration controls to countries of origin and countries of transit by the introduction of visa requirements and pre-boarding passenger checks." For example, in the early nineties, the United States used its navy to return unwanted Haitian refugees to their country, while Italy launched military operations to stop the flight of Albanians across the Adriatic Sea.[33]

Frequently, asylum-seekers are being detained in prison-like conditions, asylum criteria are being tightened, and those deemed unworthy of asylum are being deported quicker. The proportion of asylum-claimants who were actually granted refugee status by West European governments declined from 42 percent in 1984 to 7 percent in 1993 (though an additional 26 percent were allowed to stay on humanitarian and other grounds).[34]

Flows of refugees and migrants are likely to increase sharply in coming years: The centrifugal forces of war, repression, and ethnic conflict remain strong in many countries and may lead to their violent disintegration. Degradation of critical ecosystems continues to undercut the livelihoods of tens, if not hundreds, of millions of people, while the full impact of climate change is yet to be felt. And the economies of many countries may be unable to provide anywhere near an adequate number of jobs for the rapidly growing ranks of young people. Restricting the flows of people—shutting people out—will not resolve the underlying problems we face.

6

Vicious Circles: Two Case Studies

When the terrifying slaughter began in the central African country of Rwanda in 1994, many observers saw it as another example of how ethnic tensions can lead to violence. After all, this conflict seemed to pit Rwanda's majority Hutu against its Tutsi inhabitants. The sheer magnitude and ferocity of the killings seemed motivated by a near-satanic level of hatred. As horrified television viewers around the world watched bodies litter roads and pile up on riverbanks like so much driftwood, it seemed that "ancient African hatreds" must be at the root of such barbarious acts.[1]

The killings began on April 6, 1994, after President Juvénal Habyarimana, a Hutu, died in a mysterious plane crash. Many observers believe that Hutu hardliners in the army or presidential guard shot the plane down, using the incident as an excuse to kill moderate Hutu and to eradicate the Tutsi. Within hours, members of the presidential guard and pro-Habyarimana militias started a systematic, three-month slaughter during which anywhere from 500,000 to 1 million people perished. The genocide ended only when rebel forces of the predominantly Tutsi Rwandan Patriotic Front (RPF) defeated the government's forces.[2]

But what has become a routine reference to "ancient hatreds" in the post-cold war era turns out to be a poor explanation of the circumstances that brought about this genocide. Careful study shows that Rwandan society had been unraveling for several years. Rather than being a simple case of tribal bloodletting, the Rwandan apocalypse was rooted in a complex web of explosive population growth, severe land shortages, land degradation, lack of nonagricultural employment, falling export earnings, the pain of structural economic adjustment, and savage competition among the country's elites. Rwanda's authoritarian government was unable to cope with these challenges.

The distinction between Hutu and Tutsi in Rwanda rests not on biological, cultural, linguistic, or religious differences (none of which are present) but on the different origins of the two groups as cultivators and pastoralists. The Tutsi formed a warrior caste that exercised political authority and commanded the bulk of the wealth. They established a feudal system based on their control of land, cattle, and the armed forces. Yet this system also involved mutual obligations and loyalties that softened the edges of the caste system. And it provided for some mobility between the two groups.[3]

The once relatively fluid boundaries between Tutsi and Hutu became more rigid preceding colonization and particularly during German and Belgian colonial rule. Since the Tutsi benefited during the colonial period, the struggle for independence (which was achieved in 1962) was in a sense also a struggle between Hutu and Tutsi elites for state control.[4]

On the eve of independence, a key manifesto issued by a group of Hutu—while demanding an end to Tutsi rule—denounced the social and political conditions under which most Hutu farmers as well as the poorer strata of the Tutsi lived. But in an ultimately failed effort to retain its hold on power, the Tutsi elite "ethnicized" the conflict—a strategy pursued as well by the Hutu after they won power in 1961. Distinguishing between Hutu and Tutsi had, by that time, become virtually impossible due to a high rate of intermarriage; identity cards first introduced by the colonial powers somewhat arbitrarily determined a person's "ethnicity" by the paternal line of ancestry. The Tutsi increasingly faced discrimination in education and employment. The road to independence was punctuated by bouts of violence, including large-scale massacres of Tutsi in 1959 and 1964 and a subsequent exodus of many Tutsi to Uganda and other neighboring countries.[5]

By 1967, violence had ended, but the power and allure of "ethnicizing" conflicts remained, as did the discrimination against the Tutsi. In October 1990, the Rwandan Patriotic Front invaded Rwanda from Uganda. Although its fighters were primarily drawn from Tutsi who either had fled Rwanda after independence or were descendants of such refugees, it included Hutu opposed to the Habyarimana regime. The RPF's claim that it was not fighting an ethnic war was largely validated by its conduct, yet government leaders in the capital easily convinced a majority of Hutu that the invasion was an attempt to resurrect Tutsi feudal rule.[6]

By 1992, the RPF had captured a significant portion of territory in northern Rwanda, including important coffee- and tea-producing areas, the loss of which considerably reduced the government's export revenues. A shaky cease-fire was put into effect in July 1992, and a peace agreement was signed in Arusha, Tanzania, in August 1993. It provided for the establishment of a transitional government, elections, and the formation of a new national army uniting government forces and the RPF.[7]

As Valerie Percival and Thomas Homer-Dixon of the Project on Environment, Population, and Security at the University of Toronto note, however, many members of the government were intensely unhappy with the Arusha agreement because it gave the RPF control of key ministries and substantial influence in the armed forces and provided for the return of Tutsi refugees from abroad. The regime decided to block implementation of the Arusha accord, and "used every possible opportunity to increase social cleavages to create animosity toward the RPF." Anti-Tutsi feelings were much more prevalent among the Hutu elite than among ordinary Hutu; it took a massive campaign spreading fear and hatred to change that.[8]

These events took place against a backdrop of intensifying pressure in a number of areas, which helped to polarize Rwandan politics and produce the rise of extremist forces. Due to a high rate of population growth (the number of Rwandans tripled between 1954 and 1993, to 7.5 million), Rwanda became the most densely populated country in Africa. By the time conflict broke out, average farm size had declined to less than a half-hectare as plots were subdivided from one generation to another. And with half the population in 1993 younger than 15, family plots were destined to shrink even further because nonagricultural job opportunities remained limited.[9]

Because Rwanda is an overwhelmingly agricultural country—the World Bank lists it as the least urbanized country in the world, with only 6 percent of the population living in cities in 1993—the importance of land issues can hardly be overstated. With strong land pressures, much of the tension in Rwanda revolved around disputes over the distribution of pastureland and farmland. Growing landlessness caused increasing desperation and set in motion a process of social disintegration that helps explain the later descent into massive violence.[10]

Between 1970 and 1986, the total area under cultivation actually expanded by half. Much of this, however, involved marginal land with poor soil quality or inadequate rain. Losing much of their pastureland, the Tutsi suffered most from this conversion to cropland. At the same time, because a considerable portion of the new land went to coffee plantations owned by a relatively small number of people, rather than to domestic food production, many Hutu did not benefit either.[11]

By the mid-eighties, the expansion of cropland came to an end. Cultivation of food-growing areas intensified even more, and half of all farming took place on hillsides vulnerable to erosion. Overcultivation and soil erosion began to take their toll, causing land fertility to decline sharply in some parts of the country. Per capita food production took a nosedive, turning Rwanda from one of the brighter spots in sub-Saharan Africa to one of its worst within a decade. The problems were accentuated by a severe drought in 1989 that left 300,000 people dependent on food aid.[12]

Although Rwanda had been almost fully self-sufficient in cereals at the beginning of the eighties, it finished the decade having to import 21 percent of what it needed to consume. Then the closure of the agricultural frontier, declining yields, and the impact of the RPF's invasion combined to cause a virtual freefall in Rwandan food production after 1990. Grain harvests slumped from 269,000 tons in 1990 to 184,000 tons in 1993, and to 104,000 tons in 1994, the year of the genocide. (See Figure 6–1.)[13]

Parts of the country—particularly the southwest—were more vulnerable to environmental degradation than others. But farmers in the northwest fared better for an additional reason: the government channeled most of the international development aid into the region because it was President Habyarimana's home and political base. The northwest therefore was able to develop

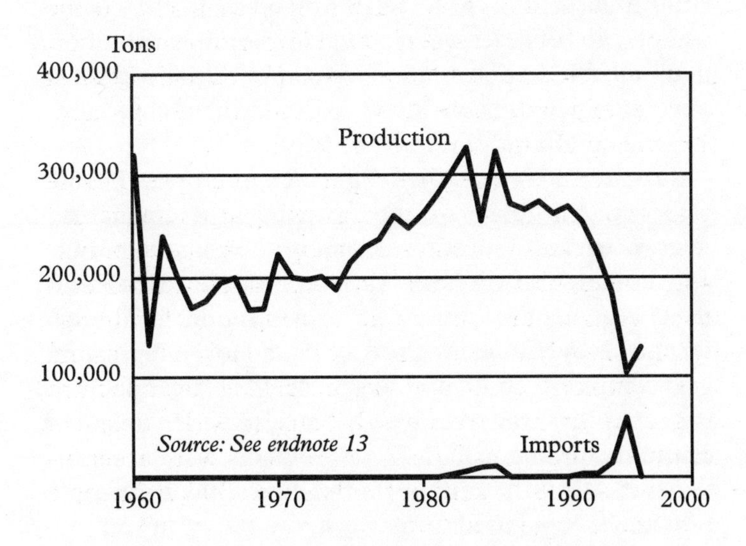

FIGURE 6–1. *Grain Production and Imports in Rwanda, 1960–95*

more, and to counter soil erosion and deforestation better, than the rest of the country. Not surprisingly, such favoritism generated resentment elsewhere.[14]

Foreign lenders had encouraged Rwanda to step up its cultivation of coffee for export as a way to finance the rapidly rising foreign debt. By the early nineties, coffee accounted for almost 70 percent of Rwanda's total export earnings. But as a tiny producer (accounting for less than 1 percent of the global coffee crop), the country has virtually no control over the market. The collapse of coffee prices during the eighties and early nineties cut revenues substantially. Rwanda's economy was especially hard hit since world market prices for tea, its second most important export crop, followed a similar trajectory. Per capita gross domestic product fell from $330 in 1989 to $200 in 1993. External debt grew from $190 million in 1980 to $910 million in 1993, when it equaled about seven years of export earnings and more than half the nation's annual income. The World Bank now ranks Rwanda as one of 32 severely indebted low-income countries.[15]

Around the time of the RPF's invasion, western lenders forced the government to accept a structural adjustment program. The severe economic crisis, the pain of adjustment, and the civil war combined to undermine the government's legitimacy. It proved harder and harder for the Habyarimana regime to reconcile competing regional interests in Rwandan society, to satisfy the elite's expectations, and to solve the accelerating social and economic problems.[16]

Rwanda's difficulties gave rise to a growing opposition movement based primarily in the southern and central portions of the country. After coming to power in a military coup in 1973, Habyarimana ruled Rwanda as a dictator. The opposition and international lenders increasingly demanded the introduction of a multiparty system. Habyarimana acceded to these demands, but maneuvered to slow the process of democratization and retain effective control. Only in 1992, following mass demonstrations by the opposition, did he agree to a multiparty transition government. At the insistence of the opposition and foreign lenders, the regime also agreed to open peace negotiations with the RPF.[17]

Hutu extremists fervently opposed these developments and gained strength within the ruling party, the National Republican Movement for Democracy and Development. The regime had already used the 1990 RPF invasion as a pretext for mass arrests. Militias were formed and they killed hundreds of suspected political opponents, both Hutu and Tutsi, who were accused of being rebel supporters. These acts of violence and intimidation were stepped up after the signing of the Arusha peace agreement in 1993. The lack of protest by the international community about these killings during 1990–93 served as an indirect encouragement to genocide. A report by Human Rights Watch points out: "Well aware of the susceptibility of foreigners to explanations of 'ancient tribal hatreds' among Africans, the government repeatedly underlined the supposedly tribal nature of the killings in an effort to mask the deliberate role played by the authorities."[18]

Preparations for the genocide were made over several months. Following Habyarimana's death in April 1994, parts of the army and the militias systematically killed members of the transitional government, proponents of peace with the RPF, and many Tutsi. The extremist Hutu militia recruited primarily the uneducated, the unemployed, and young toughs; others were forced to join in bloodshed in order to avoid being killed themselves. Despite several reports by the United Nations and independent groups in 1993 and early 1994 that provided clear and detailed warnings of the impending genocide, the international community failed to prevent the slaughter or even to stop it once the killings had begun.[19]

This brief account shows that a complex interaction of factors produced the fateful events in Rwanda. Severe economic, demographic, and environmental pressures on Rwandan society unleashed unresolved grievances, while extremist forces among the Hutu manipulated ethnic identities and resorted to large-scale violence rather than allow the transition to a multiethnic, nonauthoritarian system to take place.

* * * *

On January 1, 1994, the day the North American Free Trade Agreement (NAFTA) came into force, several hundred rebels of the Zapatista National Liberation Army (Ejército Zapatista de Liberación Nacional, or EZLN)—a previously unknown force drawn from marginalized peasant and Mayan indigenous communities—seized towns in the eastern and central parts of Chiapas in southern Mexico. In less than two weeks, government troops had forced the EZLN to retreat to inaccessible strongholds in eastern Chiapas.[20]

But the real challenge posed by the Zapatistas was of a political rather than a military nature, pointing to rampant economic inequalities and the lack of political legitimacy, and demanding land reform and democracy. Their rebellion galvanized national discussion of these issues, and although many disagreed with the EZLN's resort to violence, the rebels' demands found wide resonance throughout the country. Citizens in Chiapas and neighboring states seized dozens of town halls to protest fraudulent elections that kept the ruling party in power. Land-hungry peasants invaded private ranches; some landowners fled, while others counterattacked with hired gunmen. Under intense national and international scrutiny, Mexican President Carlos Salinas de Gortari was forced to initiate peace talks with the EZLN.[21]

Chiapas has been described as "a rich land, a poor people." With 3 percent of Mexico's population, the state accounts for 5 percent of the national production of oil, 12 percent of the natural gas, 13 percent of the corn, and 46 percent of the coffee. Even though Chiapas provides half of Mexico's hydroelectric power, only one in three households is hooked up to the electricity grid. The state is a leading producer of beef, yet fewer than half the people there regularly eat meat. In 1990, only 58 percent of households had running water, compared with 79 percent in all of Mexico; literacy was at 70 percent compared with 87 percent nationwide. Chiapas also lags behind in household income and education, and has above average rates of infant mortality.[22]

The state itself is characterized by enormous discrepancies between a relatively prosperous western region (with coffee plantations along the Pacific Ocean coast and commercial farming in the Grijalva River valley) and the impoverished eastern lowlands, which are primarily inhabited by descendants of Mayan Indians. Cut off from government services, political power, and economic opportunity, eastern Chiapas—in the words of George Collier, Professor of Anthropology at Stanford University—"is a kind of dumping ground for the marginalized." All the indicators of social and economic well-being are far lower in this region, particularly for its indigenous peasant communities. Accounting for one third of

Chiapas's population, Indians have suffered discrimination and oppression ever since the Spanish colonial conquest. Hunger and disease killed 30,000 of them during 1993.[23]

Land is one of the key issues in the Zapatista rebellion. Population growth has clearly exerted pressure: Chiapas's numbers doubled between 1970 and 1990, and the per capita amount of cultivated land has been declining since the mid-seventies.[24]

But the crucial issue—which has bedeviled the area ever since Spanish colonialism—is a highly unequal land tenure. A small number of farming and ranching elites control much of the best land in Chiapas and dominate its political system. Among coffee producers, the top 0.15 percent (those with more than 100 hectares) own 12 percent of the coffee-growing land; 91 percent of all coffee growers own less than five hectares each. Cattle ranching, dominated by a handful of *ladino* (nonindigenous) elites, has made significant inroads since 1960. An estimated 45 percent of Chiapas's territory, and a significantly larger share in the eastern portion, is now used as cattle pastureland.[25]

The rise of ranching has been observed all over Mexico since the sixties and seventies. To serve an expanding urban demand for meat, large swaths of land were turned over to cattle pasture, and a growing number of Mexican farmers switched from foodgrains to feedgrains. Particularly in southern Mexico, which includes Chiapas, many peasants "were pushed off their lands by cattlemen protected by the army and their own paramilitary bands," according to Tom Barry of the Interhemispheric Resource Center in Albuquerque, New Mexico. The consequence of this reorientation toward ranching was that by the seventies Mexico lost its self-sufficiency in grains and became a rapidly growing net importer; by the eighties, a quarter of the grain consumed there was supplied by U.S. agribusiness.[26]

In Chiapas, most land struggles are taking place in the eastern half of the state. Virtually uninhabited until the mid-twentieth century, the region—and particularly the Lacandón rain forest—attracted waves of peasants fleeing land scarcity, dislocation from dam construction, and persecution during the sixties and seventies. They were joined by Guatemalans fleeing civil war in their country. Population in the Lacandón has grown twenty-five-fold since 1960. Peasants in eastern Chiapas are locked into competition not only with each other but also with loggers and ranchers who continue to control the local governments.[27]

As growing numbers of peasants, ranchers, and loggers pressed into the area, the region suffered an enormous loss of forested lands. Tree cover in the Lacandón rain forest declined from 90 percent in 1960 to 30 percent today. The intensive cultivation practices appropriate to the climate and soils of the highlands from which most new settlers came soon exhausted the land, and the peasants were forced to move further into the forest. Meanwhile, ranchers turned forests into grazing lands—sometimes in direct competition with peasants; at other times moving in after peasants could no longer cultivate the degraded land. The amount of pastureland within the Lacandón doubled during the eighties. All told, large swaths of eastern Chiapas are now so deforested that the border between Chiapas and neighboring Guatemala is clearly visible from space. As deforestation and soil degradation march on, land pressures intensify.[28]

During the sixties and seventies, peasants throughout Mexico began to make claims on more, and higher quality, land. The ruling party (Partido Revolucionario Institucional, or PRI) has staked much of its legitimacy on the promise of land reform ever since it came to power. Indeed, George Collier notes that despite their poverty, "Chiapas peasants have been among the most reliable supporters of the ruling party since the 1930s," when President Lázaro Cárdenas redistributed a tenth of Mexico's land to peasants and indigenous communities, pushing the number of landless laborers nationwide down from 69 to 36 percent of the rural work force.[29]

During the three decades that followed the Cárdenas presidency of 1934–40, the pace of land distribution slowed dramatically, and the land turned over to small peasants was much more marginal. In Chiapas, the government distributed lands that the large landowners found least desirable, and it often promised the same piece of land to different peasant communities. Peasants' land claims typically languished in a tangle of state bureaucracies for years. More so than in other Mexican states, the large landowners of Chiapas have been able to delay or entirely resist land transfers. In 1992, unresolved land claims and unfulfilled promises in Chiapas accounted for almost 30 percent of the total land reform backlog in all of Mexico. Thus, land distribution continued to be highly unequal.[30]

Throughout southern Mexico, landless peasants and indigenous communities concluded that the only way to get land they coveted was to occupy it. Landowners responded to a growing wave of takeovers with violence. The result was an endless string of local land skirmishes, in which state and federal governments almost invariably sided with the big landowners. In the state of Guerrero, two peasant guerrilla forces were crushed by the Mexican armed forces in the early seventies—the last major guerrilla challenges to the Mexican government before the emergence of the EZLN. To defuse the situation, President Luis Echeverria, in office 1970–76, returned to a more reformist policy and initiated the colonization of Mexico's remaining frontier lands, including eastern Chiapas.[31]

Cárdenas-style populism in the seventies gave way in the eighties and nineties to technocratic, neoliberal governments that gradually abandoned the commitment to agrarian reform. In 1992, President Salinas halted land reform by amending Article 27 of the Mexican Constitution. He thus ended a government pledge that had its origins in the founding of the Mexican Republic and, although far from fulfilled, was critical to social and political stability. In effect, Salinas disavowed the peasantry as a constituency.[32]

Based on the assumption that most peasant agriculture was inefficient and an obstacle to capitalist development, the repeal of the land reform legislation was a deliberate measure intended to boost modernized, world-market-oriented farming. Collier comments that this step "robbed many peasants not just of the possibility of gaining a piece of land, but, quite simply, of hope." The formal end of land redistribution—together with the removal of agricultural tariffs under NAFTA, which presented a clear threat to the livelihoods of millions of small-scale farmers—played a key role in propelling the Zapatistas toward armed rebellion.[33]

Salinas's momentous policy change must be seen against the background of Mexico's oil boom-bust cycle during the seventies and eighties, the severe debt crisis that followed, and efforts to accelerate the country's integration into the world market. During the seventies, rising Mexican oil revenues had allowed the government to kick off ambitious development projects and to buy popular support. But Mexico also borrowed heavily, and when oil prices plummeted in the early eighties, the country found itself facing a debt burden of crisis proportions. It was forced by international lenders to trim expenditures sharply, devaluate the peso, slash wages, phase out price controls, and remove subsidies for the urban and rural poor.[34]

As credits, price supports, and subsidies diminished under austerity, rural credit went primarily to cattle ranchers; small commercial and subsistence farmers came up short. Philip Howard and Thomas Homer-Dixon of the Project on Environment, Population, and Security at the University of Toronto note that by 1990, 87 percent of agricultural producers in Chiapas had no access at all to government credit. President Salinas also dismantled INMECAFE, the Mexican Coffee Institute, which had guaranteed markets for small coffee growers in Chiapas. At a time of steeply falling world market prices, many small producers were wiped out.[35]

In 1989, the Salinas government launched a National Solidarity Program (PRONASOL). By 1994, some $15 billion had been channeled into its anti-poverty programs. Yet PRONASOL failed to win popular support in Chiapas, even though the state received more funds from it than any other Mexican state. According to Tom Barry of the Inter-hemispheric Resource Center, PRONASOL was not designed "to address seriously the rural development crisis. Rather, [it] served only to take the edge off the crisis." Although the World Bank, which supports the program, describes it a success, PRONASOL funds are being disbursed in a manner that reinforces PRI's favoritism.[36]

Despite PRONASOL, austerity eroded the PRI's overall ability to maintain its elaborate system of patronage and, hence, its political power. Instead of buying peasant loyalty, the central and state governments and the despised *caciques* (local PRI bosses) turned to more coercive measures, including denying government services to those whose political allegiance was in question or evicting people from communally held land.[37]

The eighties and nineties were marked by increased repression, including a 1980 army massacre in the village of Golonchan of peasant families who had taken over a powerful rancher's land. The 1988 presidential election was closely contested, and the PRI's victory widely considered to be due only to massive fraud. The experience of grassroots peasants' organizations demonstrated the frustrating limits of working within an authoritarian and unresponsive system as much as it showed the dangers of being co-opted and marginalized politically.[38]

Mexico's wrenching, boom-bust cycle of development brought with it important changes in Chiapas. The differentials of wealth and power within communities rose sharply. The growing gap between rich and poor and the replacement of in-kind exchanges with cash transactions broke down many traditional relationships of mutual dependency that once helped poorer citizens. Tensions grew. As a result, George Collier explains, "the wealthy no longer have a need for webs of reciprocal obligations with the poor; the rank and file whose support they once cultivated have become expendable." These changes made themselves felt just when government services were slashed. Hardest hit were communities in the already poor eastern part of Chiapas. Collier concludes that "the impoverished have had no place to turn and little to lose by joining the Zapatista rebellion."[39]

Although the EZLN's actions were prompted by local conditions, the insurgents "exposed the multiple failures of reform in Mexico," as summed up by Andrew Reding of the World Policy Institute: "exclusion of large segments of the population from economic development; a growing gap between rich and poor; denial of basic political rights...; brutal repression of dissent...; and the persistence of racist attitudes" toward the country's sizable indigenous population.[40]

Chiapas is in some ways an extreme case, but the conditions recited by Reding, plus worsening environmental scarcities, can be found throughout much of Mexico. Those embracing the EZLN were primarily members of a spreading underclass, but the reservoir of resentment extends beyond the marginalized. As Tom Barry points out, for example, the government's neoliberal policies also squeezed medium-scale farmers (those with 10–40 hectares) hard.[41]

Chiapas is symbolic of the problems that Mexico faces in the sense that the conflict in the state reveals the costs of a world market integration strategy that comes at the expense of a large share of the population. In this sense, Chiapas also points up challenges faced by other developing countries that have embarked on similar policies.

Peace talks between the Zapatistas and the Mexican government began on February 21, 1994, but did not make much headway. In the grip of a severe new financial crisis, and pressed by some domestic businesses and foreign investors that regarded the Zapatistas as a threat to their interests, President Ernesto Zedillo attempted a military solution to the crisis in February 1995. Troops recaptured some of the Zapatistas' jungle strongholds, but failed to stamp out their political challenge. Negotiations resumed, and one year later produced a first agreement promising limited autonomy to the 7 million people in indigenous communities throughout Mexico.

The agreement concludes the first of six separate sets of negotiations. Some of the most difficult issues, including land distribution, remain to be tackled, and the government-EZLN dialogue reached an impasse in mid-1996.[42]

* * * *

Rwanda and Chiapas demonstrate the complex interactions of several stress factors. Population growth, environmental decline, the lack of economic opportunities, and the absence of representative forms of governance are common to both cases, yet the dynamics arising out of the particular circumstances of history and political culture in each region produced sharply different outcomes. Although careful analysis of underlying stresses may not make their consequences predictable—there is no blueprint that would permit a precise forecast of genocide, low-intensity violence, or an entirely different outcome—such attention would allow policymakers to address the elements that lead to or boost societal fractures and hence to prevent disintegration and large-scale violence.

Some observers may be tempted to regard the events in Rwanda and Chiapas as having little relevance for the rest of the world. By and large, these regions are not of any "strategic" concern, and they involve marginalized and poor populations. But the same pressures can be found in innumerable communities and societies the world over, including larger ones that the international community would be hard-pressed to ignore.

Several big countries are experiencing acute and chronic fissures, including India, where ethnic-religious antipathies and challenges to a nonsectarian society are strong; Brazil, where extreme social inequalities produce explosive conditions; Zaire, where a tiny elite's looting of the country's resources have left behind an economically and politically depleted society; and China, where regional disparities and tensions are rising and the contradiction between economic liberalization and political authoritarianism produces a potentially combustible situation. Immense and growing problems can also be found in countries as different as Russia—shaken by the trepidations of an enormously difficult political and economic transition—and the United States, where rising inequality and job insecurity are increasingly leading to social polarization.

In virtually all these countries, decision makers and ordinary persons alike will be much more challenged by the issues referred to here than they are by the traditional concern of a foreign invasion. All are confronted by issues that go to the very heart of their well-being and security. But military might—the traditional answer to societal challenges—is practically irrelevant. A new security policy needs to strike a new balance between traditional national security concerns and those of human security.

II

Security in the Twenty-First Century

7

A Human Security Policy

> We share the conviction that social development and social justice are indispensable for the achievement and maintenance of peace and security within and among our nations. In turn, social development and social justice cannot be attained in the absence of peace and security or in the absence of respect for all human rights and fundamental freedoms.[1]
>
> Copenhagen Declaration on Social Development, World Social Summit, March 1995

Our contemporary world often seems more attuned to outward symptoms than to root causes, more enamored with technical fixes than with social change. Policymakers typically approach issues as though they can be treated in isolation from their context. But experience invariably contradicts this. To address rapid population growth, for instance, it is not enough to ensure a steady supply of contraceptives; rather, factors such as women's status in society, cultural values and constraints, and couples' desires to ensure their own social security through their offspring need to be taken into account. Likewise, in transportation policy the challenge is not to construct more highways to accommodate the incessant growth of automobile fleets, but to adopt land use policies and other measures that will reduce travel distances and hence reliance on motorized travel.

The same is true for security policy. Providing human security is less about procuring arms and deploying troops than it is about strengthening the social and environmental fabric of societies and improving their governance. To avoid the instability and breakdown that occurred in Rwanda and Chiapas and countless similar areas, a new security policy—one focused on human security—must take into account a complex web of social, economic, environmental, and other factors.

Depending on the circumstances, a human security policy may encompass such seemingly disparate concerns as redistribution of wealth, debt relief, job creation, technology development, more democratic and accountable governance, and the strengthening of civil society. But these concerns are closely interrelated. Addressing one set of factors while neglecting others is unlikely to yield any lasting solutions. For example, severe social inequities often trigger events that have a detrimental impact on the environment, just as environmental decline has important repercussions on social development.

There is much that national governments can do on their own to promote human security. But as important as these efforts are, human existence is increasingly being shaped by global and local trends. National policies need to be complemented by improved international cooperation among countries and by a strengthening of civil society everywhere.

It seems clear that the international community is beginning to understand the interconnectedness of the issues at hand. Several U.N. conferences during the nineties have sketched the scope and dimensions of a worldwide human security policy. Historic meetings took place on environment and development (Rio de Janeiro, 1992), human rights (Vienna, 1993), population and development (Cairo, 1994), social development (Copenhagen, 1995), women (Beijing, 1995), and the urban habitat (Istanbul, 1996). The relevance of the selected topics to peace and security was stated quite explicitly in various conference documents. The gatherings attracted not only high-level policymakers but also large numbers of grassroots and advocacy groups and legions of journalists. By generating new areas of global consensus, setting goals to be met, and soliciting commitments to meet these goals, each of these conferences helped advance the human security agenda. At the same time, they highlighted the gaps and shortcomings in current policies.

Meanwhile, the number of community, grassroots, and advocacy groups—usually lumped together as nongovernmental organizations (NGOs)—continues to grow spectacularly. Their presence and influence in individual countries varies enormously. In some countries, NGOs are well established, attract substantial funding, and help shape policy; in others, people organize in self-help groups to overcome governmental neglect or in a desperate bid to defend themselves against abuses by powerful governments and corporations. But what all NGOs have in common is that they represent a new force in decision making separate from the state and from private capital.

Because globalization and localization have both positive and negative ramifications, a human security policy needs to build on their constructive features while minimizing the disruptive ones. At the global level, improved cooperation and a new North-South bargain on social and environmental issues are crucial. At the local level, revitalizing communities and enhancing civil and human rights are essential. Achieving human security requires not only that global and local governance are improved, but that interaction between them be harmonized.

* * * *

Globalization is rapidly turning collaboration among countries from something optional into a necessity. There is perhaps no better illustration of the imperative of international cooperation than the state of the environment. Because environmental degradation respects no human-drawn boundaries, purely local or national efforts are condemned to failure. "Environmental diplomacy" to establish common goals, standards, and procedures—through such varied endeavors such as multilateral conferences, global treaty-negotiating sessions, regional forums such as joint river and watershed commissions, and bilateral meetings—is now firmly established on the international stage.[2]

During the past quarter-century, the number of international environmental treaties has skyrocketed—from about 50 in 1969 to 173 in 1994. Including less-binding types of accords and bilateral treaties, the number of international environmental instruments reaches almost 900. Still, progress is not measured just by the number of agreements on the books. The impact and effectiveness of these treaties vary enormously. The 1987 Montreal Protocol on the Depletion of the Ozone Layer, for example, has resulted in a 77-percent reduction of chlorofluorocarbon (CFC) production from 1988 to 1995. But many other accords, including the Framework Convention on Climate Change that entered into force in 1994, commit the signatories to little that is concrete. And monitoring, review, and enforcement of treaty provisions are often weak or nonexistent.[3]

The emergence of global environmental norms and standards is an agonizingly slow process. The international environmental conference circuit makes for busy schedules and churns out large volumes of documents. Yet the gap between rhetoric and action is growing. Four years after the auspicious Rio Earth Summit, international environmental diplomacy seems to have run out of steam due to a lack of political will. "The sense of a 'new dawn' that marked Rio has since been superseded by standstill and backsliding," the German NGO Forum on Environment and Development noted in May 1996. The fourth session of the U.N.'s Commission on Sustainable Development, in April 1996, was widely judged to have missed opportunities to make further progress.[4]

In the face of transboundary environmental challenges, transnational cooperation has become essential before political realities fully allow it. One obstacle is that although no country can escape the consequences of environmental breakdown, national interests are not necessarily convergent: neither the harm of environmental degradation nor the benefits arising from protective measures are equally distributed. In addition, the ability of individual communities and countries to shoulder the financial burden of preventive or remedial policies may differ considerably. Finally, environmental awareness—among policymakers and the larger population—has developed much earlier and more strongly in some countries than in others.[5]

One of the big, unaddressed tasks is to determine each country's responsibility for reducing environmentally detrimental activities. Devising an acceptable formula, for instance, to allocate among countries the massive reductions in emissions of greenhouse gases needed to avert global climate change is proving to be intensely contentious. Older industrial countries have long been the principal contributors of carbon dioxide and other gases implicated in climate change. Yet the world is now confronted with the specter of China, India, and other populous countries undergoing rapid, resource-intensive industrialization. China, in particular, is relying heavily on coal to fuel its economic progress, and may soon become the single largest source of carbon emissions. Industrial countries warn that the efforts of China and others to emulate western affluence will unhinge the global climate balance. This is a valid concern, yet the countries pointing the finger at the Third World have with few exceptions made little progress in cutting their own emissions.

Whether the concern is climate change or other environmental dangers, industrial countries are responsible for the preponderance of the degrading activity, and so should shoulder a commensurate share of the responsibility for change. They consume 80 percent or more of the world's aluminum, paper, and iron and steel; 75 percent of its energy; 75 percent of its fish catch; 70 percent of its CFCs; and 61 percent of its meat. Hence, they are responsible for the bulk of hazardous wastes generated in mining and smelting, for the clear-cutting of forests, for the air pollution and buildup of greenhouse gases associated with fossil fuel burning, for overfishing, and for a large share of the severe soil erosion found on grazing lands.[6]

Many of these degrading activities do not occur in industrial countries, but in developing ones. As Worldwatch researcher Aaron Sachs writes, "just as in colonial times, poor nations end up despoiling their own lands in order to export certain products that feed the richer nations' habit of overconsumption." For example, none of the copper extracted by the Panguna mine in Bougainville (see Chapter 3) went to local consumers. Bougainville is typical of the global pattern: copper production takes place mostly in developing countries, but per capita copper consumption is roughly 20 times higher in the industrial world.[7]

The years ahead will put environmental diplomacy to a severe test. Although the traditional treaty-making process will clearly continue to play a role, it is too cumbersome and inflexible, and tends to yield least-common-denominator results. There are a variety of time-tested ways—incentives, differential obligations, and other measures—to raise the denominator and make stricter standards more acceptable. In addition, informal "environmental alliances" among selected countries that are prepared to accept stricter goals and standards play an important trail-blazing role. By demonstrating the feasibility and acceptability of strict standards, the trail-blazers in effect invite others to join in a bandwagon effect. By repeating this process over time, the result is a ratcheting up of agreed standards.[8]

This is exactly what happened in Europe regarding air pollution control efforts, for example. A handful of countries were joined by a growing number of others in the mid- to late eighties in a "30-Percent Club." Aware that vast amounts of airborne pollutants drift across borders, these nations committed themselves to reducing their sulfur and nitrogen oxide emissions by at least 30 percent over a specified number of years. Several countries achieved the goal early and went on to reduce their emissions even further.[9]

Despite, or perhaps because of, the flurry of environmental diplomacy activities, there is a need for strategic direction. To make sustainable development more than a widely repeated slogan and to shore up the environmental basis of international security, a North-South bargain is needed. Industrial and developing countries alike will need to commit to a climate stabilization strategy that taps into the large reservoir of energy efficiency and non-fossil-fuel energy technologies. In return for abandoning a course that simply seeks to replicate the western resource- and pollution-intensive pattern of industrialization, developing countries should receive adequate financial support for programs that help avert climate change, counter land degradation, and address water scarcity. Such a bargain would not only have environmental benefits, it would also carry social and political advantages—helping to avoid the social dislocations and conflicts that now go hand in hand with environmental destruction.

★ ★ ★ ★

The other part of a North-South compact concerns social development. There is an urgent need for a new commitment to end deep inequities that breed explosive social conditions, fuel ethnic antagonisms, and drive environmental decline. The World Social Summit in March 1995 recognized that poverty, unemployment, and social disintegration are closely linked to issues of peace and security.[10]

The gathering concluded with a series of important commitments by national governments. For the first time, political leaders pledged to a global audience to prepare national strategies to eradicate poverty "in the shortest possible time." They made commitments to promote the goal of full employment and to enable "all men and women to attain secure and sustainable livelihoods." The participants understood that although a change in national priorities is important to alleviate social disparities, this needs to be accompanied by changes on the global level.[11]

Among other proposals, the heads of government gathered in Copenhagen endorsed limits on the social repercussions of structural adjustment programs. An overhaul of these programs is badly needed. It is high time to abandon the blind one-fits-all prescription that encourages developing countries to pursue an export-promotion strategy that drives commodity prices down, fails to enhance their indigenous economic capacities, and does nothing to reduce their extreme vulnerability to the world market.

Academics have long discussed the pros and cons of world market integration, recognizing that strategies and responses depend on a country's particular set of strengths and weaknesses, and that these in turn vary over time. Yet the orthodox prescription that the Bretton Woods institutions—the World Bank and International Monetary Fund (IMF)—continue to impose through structural adjustment programs insists that a virtually unconditional embrace of international market forces is the best possible path.

The economic model underlying this prescription, according to the U.N. Research Institute for Social Development, "corresponded to no known place on earth—not to any of the industrial countries of the North, and certainly not to any of the developing countries." Newly industrializing countries such as South Korea, just like older industrial countries before them, relied heavily on state intervention and protected markets, building strong and coherent economies before opening up to the world. The Bretton Woods institutions' prescription thus flies in the face of historical experience.[12]

A new adjustment policy needs to adopt a less rigid approach to market forces and world market integration and to recognize that government intervention, in proper doses and with suitable timing, is not a hindrance to economic development but an essential ingredient. It needs to focus on the most effective ways of enhancing developing countries' indigenous capacities. Depending on each country's particular circumstances, this will involve a different mix of state intervention, private enterprise, and community-based action. No single formula fits all. A North-South bargain would simply try to create an enabling environment and to remove current obstacles.

One of the key measures in this context is far-reaching debt relief. Efforts by debtor nations to service their foreign loans have hemorrhaged their economies and increased the strain on an often already frayed social fabric. Stepped-up exports of oil, timber, minerals, and other natural resources in order to generate foreign exchange revenues for debt service have also led to increased pressure on indigenous peoples in resource-rich lands, evictions of small-scale farmers and pastoralists by commercial agribusiness, and more environmental stress. And yet, huge debt payments have not gotten these countries out of the debt trap. Mexico, for instance, has transferred more than $150 billion to its public-sector creditors over the past 15 years, but it still owes them $100 billion.[13]

So far, bilateral creditors have granted some modest debt relief; multilateral creditors such as the World Bank and the IMF, meanwhile, have not cancelled any debts. Yet debt service payments flowing to multilateral creditors are increasing in importance: the amount of money involved tripled between 1980 and 1994, and the share of total debt service payments grew from 20 to 50 percent. Prodded by NGOs such as Oxfam International, the World Bank has begun to acknowledge that the debt burden of the poorest countries is essentially unmanageable and that existing debt policies are inadequate. The Bank and the IMF (though much less enthusiastically) are reconsidering their approach.[14]

A framework paper produced by the Bank and the IMF insists that debtors deemed eligible will receive debt relief only after they have completed five years' worth of additional structural adjustment programs to the satisfaction of the Bretton Woods institutions. Hence, the proposal would impose uncertainty and more pain before providing any real relief. Implementation of the plan was blocked in early 1996 by disagreements over the burden that the western industrial countries would be asked to shoulder in order to finance a new debt relief fund. Further deliberations are now expected to take place during the annual meetings of the World Bank and the IMF in October 1996.[15]

Real progress, however, requires a speedy resolution of the issue. Bilateral and multilateral creditors need to reduce the crushing burden of debt for the world's poorest countries. Failure to do so—pretending that these debts can indeed be paid—may look good in creditors' account ledgers, but it will continue to contribute to instability and social upheaval in debtor countries, which in turn may require substantial infusions of emergency spending further down the road. It is time to adopt a far-reaching debt cancellation policy, in particular for the weaker and more vulnerable developing countries.

★ ★ ★ ★

The inequities that threaten social stability and environmental sustainability extend beyond the North-South dividing line. They exist on several other levels—between men and women, between rural and urban populations, and between rich and poor within individual countries. The severe maldistribution of land, for example, will continue to push the landless into rain forests and onto ever steeper hillsides, fuel land wars, and drive the dispossessed into teeming urban slums. (See also Chapter 4.) In the cities, enormous disparities in wealth and welfare will continue to feed resentment, producing an explosive mix of anger and desperation.

In some countries, such as Brazil, so much land is left idle by absentee landlords that redistributing it would virtually eliminate landlessness or make a huge dent in it. But land reform—redistributing land, guaranteeing secure land tenure, and improving the availability of credit and extension services—has additional benefits, as the experience of South Korea, China, and some other East Asian countries has shown. Comprehensive land reform there led to higher rural incomes, transforming what otherwise would be persistently unmet human needs into effective market demand. This in turn helped stimulate industries, which provided much-needed employment outside agriculture. It was the resulting higher social and economic security that in turn helped restrain population growth.[16]

But in other countries, particularly in Latin America, land reform has languished. Efforts by the landless or near-landless to take over large landowners' estates have been met by wholesale repression in Central America; they have caused skirmishes in Chiapas and elsewhere in Mexico; and they continue to trigger bloody confrontations in Brazil, where 1 percent of the population controls 45 percent of the land. Events in Brazil are clear testimony to the scale of the difficulty in overcoming the interests of the rural oligarchy. Under President Fernando Henrique Cardoso, the Brazilian government has committed itself to a modest degree of land redistribution; yet Cardoso is being severely tested by landowners, some of whom have hired private armies to retain control, and by the local and state governments dominated by the latifundistas. According to the Movimiento dos Trabalhadores Sem-Terra, some 85,000 landless families are currently involved in land occupations; more than 1,000 persons have been killed in confrontations with landowners in Brazil during the past 10 years.[17]

Far-reaching redistribution of land would go a long way toward stabilizing rural communities. Another key measure is the provision of credit, and there is indeed a growing movement to extend "business" credit to the rural and urban poor who could never hope to receive loans from commercial banks for lack of sufficient collateral and who are otherwise bypassed by the global economy. These micro-loans can help individuals and communities escape the poverty trap, and can generate jobs and income that will help communities gain a more secure footing so that they can again be anchors of society instead of sources of migrants and pools of festering resentment.[18]

Perhaps the best known example of micro-loans is the Grameen Bank in Bangladesh, set up by Mohammad Yunus in 1983. Yunus began lending small amounts of money to impoverished villagers—mostly women, and at a fraction of the interest charged by moneylenders—so they could start up micro-enterprises in rice processing, rickshaw transportation, weaving, and other areas. "Access to credit should be a human right, irrespective of economic situation," says Yunus.[19]

Groups of borrowers are asked to accept mutual liability for the repayment of loans: if one defaults, others have to make payments. This peer pressure system works extremely well, resulting in payback rates much higher than those achieved by commercial banks. And villages in which the Bank operates have prospered more than those where it was not active. As one measure of its success, the Grameen Bank reached a total loan volume of almost $500 million in 1995, making it the world's largest financial institution serving the poor. It has made loans to 2 million Bangladeshi families in 35,000 villages.[20]

The Grameen Bank has inspired similar initiatives in countries around the world—from Guatemala to Ghana to Indonesia. The number of people served by micro-credit institutions—banks and cooperatives—has risen from 1 million in 1985 to about 10 million now. But this is still only a small percentage of those who are being denied credit by commercial banks. There are plans to convene a "micro-credit summit" in early 1997, in an effort to reach as many as 100 million of the world's poorest families by 2005. Even the World Bank, more known for its involvement in gargantuan projects, has now pledged some $200 million for micro-enterprises.[21]

One of the core ideas inherent in the micro-credit loans—a strong commitment to community—is relevant to more than just the poor and disenfranchised. In a global economy ruled increasingly by corporations that are less and less bound or loyal to any particular location, the fate of communities everywhere hangs in the balance. The dynamics of global capitalism are such that corporate management pays more and more attention to improving the bottom line for the next quarter than it does to the long-term stability of communities that it operates in or sells its wares to. Hence, applying labor-saving technologies—getting rid of workers—is now a top goal of management. To provide stability of jobs and incomes and to ensure adequate reinvestment of profits in the communities where they were generated, there is a growing need to put more emphasis on community-centered development.

Governments virtually everywhere have increasingly turned to privatization and deregulation during the eighties and nineties. Deregulation is taking place against the backdrop of rapidly expanding globalization and capital mobility. Without global standards that set the parameters for business operations, communities everywhere will increasingly be compelled to compete directly with each other for the jobs and investments that corporations may offer in return for a "friendly" investment climate.

The outcome of such a free-for-all is likely to be a low-common-denominator world with regard to wage levels, working conditions, civil rights, social welfare, and environmental laws. In a globalizing world, strong international standards—a high-common-denominator approach—are needed to prevent a race to the bottom. In some areas, such as human rights, standards exist but need better enforcement; in others, adequate standards are yet to be agreed on. A comprehensive code of conduct for corporations, for example, has been made into a virtual taboo subject by western opposition.

Of course, individual companies or industry associations have adopted some codes of conduct, such as the chemical industry's Responsible Care program. But the standards adopted are often highly uneven, adherence is partial and essentially voluntary (although community or investor pressure may reduce the degree of voluntarism involved), and implementation hinges on self-policing. Corporations' own codes may evolve into broadly accepted operating principles, and there is a continuum rather than a sharp distinction between voluntary codes and legislatively imposed standards. But what is important about any code is that it reflect the demands of affected communities, that its application be universal, and that corporations can be held accountable. The world is still a far way off from that.[22]

A new compact between communities and corporations would reward companies that engage in good "citizenship"—that score well on environmental and health issues, job creation and labor relations, and their overall relationship with host communities. It is worth remembering that corporate charters are not a right, but a privilege that can be revoked if a company continuously and massively fails to adhere to international standards.[23]

★ ★ ★ ★

Safeguarding and enhancing the rights of citizens and communities—vis-à-vis both governments and corporations—is an integral component of a human security policy. Included here are traditional human rights and civil liberties such as freedom of expression and protection against government repression, but also measures to ensure access to relevant information, to make corporate and bureaucratic decision making more transparent and accountable, and to improve community participation.[24]

At a minimum, such rights will give communities tools to defend themselves against projects that imperil their well-being and existence, such as the Panguna copper mine in Bougainville or the Sardar-Sarovar dam in India. (See Chapter 3.) But beyond purely defensive purposes lies a much greater realm: the active involvement of local communities in shaping development projects or other economic endeavors. Given the opportunity, local communities are often strong promoters of sustainability. This, at least, is what the experience of such varied endeavors as the Sangam organic farming project in Andhra Pradesh, India, and the Yanesha sustainable forestry cooperative in the Peruvian Amazon suggest. As Aaron Sachs has observed: "Over and over again, participatory programs have proven to be excellent tools in the promotion of environmentally sound, equitable development."[25]

A key way to increase accountability is to improve the flow of information. The United States pioneered so-called community right-to-know legislation in the eighties, and some other countries have adopted similar laws. U.S. legislation requires that governments and corporations furnish a broad range of environmental data; under the Toxics Release Inventory (TRI), for example, U.S. firms must report releases of some 300 toxic chemicals. Thanks to this, citizen groups have been able to better monitor environmentally harmful practices and to put pressure on polluters to improve their operations.[26]

As useful as the TRI is, communities need even broader access to information about operations that affect their quality of life and their future. This could be accomplished by corporate environmental and social audits that provide a more comprehensive picture of the impacts of operations to the communities in which they operate or intend to operate.

Although the struggle for human and civil rights, for transparency and accountability, is far from won, there is now an appreciable rise of civil society worldwide. The number of NGOs—grassroots activists, community associations, labor unions, human rights watchdogs, and pressure and advocacy groups of all stripes—is rising rapidly. NGOs are making inroads virtually everywhere, albeit often under threat of harassment, censorship, and violence in more repressive countries. And the phenomenon is not limited to individual nations: the independent Commission on Global Governance noted that the number of international NGOs (defined as those operating in at least three countries) grew from 176 in 1909 to 28,900 in 1993. By 1994, more than 1,300 NGOs (not counting national U.N. associations) had "associative status" with the United Nations, and more than 1,400 were officially accredited to the Earth Summit in 1992.[27]

With the rise of NGOs (or, as some now prefer, CSOs—civil society organizations), decision making in matters ranging from development and environment to peace and security is being transformed. No longer can governments engage in secret diplomacy against their own people, and no longer can corporations easily hide behind a smokescreen of proprietary information and private property rights. This does not mean that the outcome will always or even in most cases be favorable to NGOs, but rather that the interests of the communities they represent will have to be taken into account much more than was the case previously. Indeed, locally, nationally, and globally, NGOs are increasingly involved in influencing the policy agenda. Although some NGOs focus on relatively narrow concerns and issues, and their own internal decision making may not always be democratic, the proliferation of such groups does help to make societies more pluralistic and representative.

Recent U.N. conferences on environment, social development, and other topics were landmark events in part because of the very active and vocal involvement of NGOs. But the NGO role is equally crucial in the follow-up to these gatherings—the effort to hold governments to their promises and to monitor progress in implementing conference action programs. As the experience with the post-Rio process, and particularly with regard to national reports on implementing Agenda 21, demonstrates, continued grassroots pressure is critical for any real progress to occur. The same is true for the other conferences. To monitor progress by individual countries after the World Social Summit, for instance, NGOs set up a Social Watch project.[28]

To remain relevant and effective, the United Nations will need to become more of an organization of peoples, not merely of state representatives. In its report *Our Global Neighborhood,* the Commission on Global Governance recommended that an annual Forum of Civil Society be convened at the United Nations, consisting of representatives of organizations accredited to the world body, and that it be supplemented by regional forums that would enable a wider number of NGOs to provide input to the global forum.[29]

Ironically, in this era of privatization and anti-government sentiments, there remains a strong need for government at the local, national, and global levels—albeit government more responsive to pressing social and environmental needs, more attentive to the repercussions of globalization, and more imaginative in pursuing partnerships with a wide variety of civil society organizations. But governance is too important to leave to governments alone.

8

Enhancing International Peace Capacity

An old adage asserts that the generals are always refighting the last war—preparing for conflicts with the assumptions and methods of old, and inattentive to changed circumstances. But in today's world, the challenge is not to avoid refighting the last war, but to avoid fighting wars altogether. In the conventional view of security, as described in Chapter 1, military power plays a central role. Yet from the perspective of human security—safety from repression, social well-being, and protection from environmental degradation—the utility of military power is much more circumscribed.

Three reasons for this stand out. First, although there are undeniably cases in which one state feels threatened by another (for example, Pakistan and India, or Ecuador and Peru), conflicts today are predominantly of an internal nature. The military is often used as an instrument of domestic violence and repression. In 1992, according to Ruth Sivard, 94 developing countries suffered repression of internal dissent, which was particularly harsh among the 61 countries under military control.[1]

Second, the pursuit of military might is such a costly endeavor that it drains away resources urgently needed to meet people's social needs in areas such as food, housing, education, and health care. The military consumes resources that could help reduce the potential for conflict within societies where many of these basic needs are barely, if at all, fulfilled. (See Chapter 4.)

Third, it is difficult to see how the military could enhance the environmental foundations of security. Military power cannot reverse deforestation or soil erosion. Tanks and warplanes cannot repel the airborne and waterborne pollutants that emanate from domestic sources or cross borders with impunity. Actually, the flow of pollutants across borders is a poor analogy to military assault: the degradation of natural systems is not the work of a determined aggressor but rather the product of a multitude of routine (though flawed) economic activities. Not only do military means contribute nothing to achieving environmental security, they detract from it in a variety of ways. Modern warfare entails large-scale environmental destruction; peacetime operations—the production and testing of weapons, the conduct of maneuvers, and the generation of military-specific wastes—have a considerable environmental impact.[2]

In short, the military is an inappropriate and often detrimental factor in a broader view of what constitutes security. The military is a depreciating asset. It is true that members of the armed forces have repeatedly been recruited for humanitarian assistance and other nonmilitary tasks. And there are efforts to use some of the military's assets—from underwater sonars to space satellites—to monitor and assess the state of global ecosystems. To the extent that the armed forces are used for such purposes, however, it is their access to resources that makes them useful. It would be far better to create adequately equipped institutions that are fully dedicated to tasks that the military is called on occasionally to perform.[3]

★ ★ ★ ★

The security challenges of our time require not a recalibration of military tools and strategies or the reassignment of the armed forces, but a commitment to far-reaching demilitarization. Conflicts typical of the contemporary world cannot be resolved at gunpoint. Balance-of-power and deterrence prescriptions hardly ever work between nations, let alone within them.

Governments and individuals alike may feel that an abundance of weapons provides a stability and security of sorts, but over time the result may be the exact opposite. The immense worldwide proliferation of arms of all calibers—from assault rifles to howitzers to jet aircraft—makes it more likely that conflicts will be carried out by violent means, that fighting will be of higher intensity and last longer, that greater havoc and destruction will be wrought in the process, and that peace—after fighting comes to an end—will rest on shaky foundations.[4]

Arms are all too easily available through a variety of channels. The first source is the arms trade itself, including government-sanctioned sales, covert supplies, and gray and black market deals. Much attention is focused on exports of sophisticated arms such as jet fighters or battle tanks, but enormous numbers of smaller, less advanced arms have been transferred as well. For example, since World War II, some 55 million AK47 and AK74 Kalashnikov automatic rifles have been produced, many of which were exported.[5]

Second, surplus arms from arsenals in the North Atlantic Treaty Organization (NATO) and former Warsaw Pact countries are being sold off. The armed forces of several countries are holding virtual fire sales to get rid of surplus stocks. Germany, for example, has sold naval ships and large amounts of ammunition left over from the defunct East German army. In the former Soviet Union, theft of weaponry from the enormous Soviet inventory is on the rise, and the Russian armed forces have emerged as a new supplier, at a time when porous borders and ineffectual government authority weakened export controls. In all, secondhand sales accounted for 30–40 percent of global major arms exports in 1994.[6]

Third, surplus arms are being transferred from one Third World "hot spot" to another. The end of wars in Central America and in parts of Asia and Africa resulted in large stockpiles of surplus arms—mostly "battle-proven" small arms. Some of these weapons remain in the hands of former combatants who, disgruntled by their poor prospects in peacetime, may reemerge as a new fighting force or turn to criminal behavior. Some weapons are smuggled into other countries.[7]

And fourth, a growing number of countries are producing a range of weapons domestically. Arms export transfer deals now often include a shift of production capacities or know-how, leading to an increase in the number of producers and suppliers. Worldwide licensed production of major conventional weapons systems, measured by the number of different systems produced, increased from about 20 in the mid-sixties to more than 180 by the early eighties, and has remained high ever since. Half these licensed systems are being produced in developing countries. The capacity to produce smaller-caliber arms is even more widespread. All in all, the 1993/94 edition of *Jane's Infantry Weapons* reported that some 1,700 different weapons are being produced by 252 manufacturers in 69 countries.[8]

Publicly known exports of major weapons climbed to more than $70 billion in the mid-eighties, but fell to about $20–25 billion in the early nineties. But these figures do not include small-arms transactions (estimated at anywhere from $2–10 billion per year), and they suffer from other deficiencies. And it is far from certain that the decline of the past few years is more than a temporary phenomenon. Without any real constraints in place, sales may well increase again in the future. In a February 1995 study, the Pentagon projected worldwide sales in just seven major weapons categories during 1994–2000 at more than $200 billion; the U.S. Defense Intelligence Agency, meanwhile, estimates the global arms market for 1991–2000 to be on the order of $334 billion. If the latter figure proves correct, it would mean average annual sales roughly in line with the period from 1960 to 1990, when about $1.2 trillion worth of arms were sold.[9]

Governments justify their arms sales by arguing that they contribute to legitimate defense needs and regional stability. The Clinton administration, for instance, insists that it sells only to "responsible" parties. But the evidence indicates otherwise. An analysis by William Hartung of the World Policy Institute shows that the United States supplied arms to one or more protagonists involved in 90 percent of 50 major conflicts going on in 1993–94. In 36 percent of these, U.S. supplies accounted for a quarter or more of all arms deliveries; in 12 percent, they accounted for three quarters or more.

Often the deliveries continued even after fighting had broken out. "The decision to sell is made based on short-term political, strategic, or economic considerations," according to Hartung, "with little thought given to how these arms might be used a few years down the road."[10]

The U.S. supply of arms to the Afghan Mujahideen during the eighties provides a good example of the destabilizing consequences of such policies. In response to the 1979 Soviet invasion, millions of tons of military materiel—precise amounts are unknown due to secrecy and poor recordkeeping—were pumped into the region through Pakistan by the Central Intelligence Agency. U.S.-supplied weapons sustained the resistance to the Soviet occupation, and later fueled a ferocious, ongoing civil war among competing Mujahideen factions after the Soviet withdrawal in 1989. And the arms "pipeline" had significant leaks. Afghan rebels and Pakistani intelligence officers diverted large amounts of military equipment for resale. Weapons from the Afghan pipeline turned Pakistan's North West Frontier Province into a massive arms bazaar and fueled a dramatic increase in violence in Sindh Province. They have also been smuggled into civil war–plagued Tajikistan, India's Punjab region, and Kashmir, where they increased the severity of the violence between Indian forces and proindependence militants.[11]

Unfortunately, the story of the leaked Afghanistan arms pipeline is far from an aberration. Weapons left behind by the United States in Vietnam in the seventies showed up in the Middle East and Central America; U.S. armaments pumped into Central America in the eighties are now part of a regional black market where Colombian guerrilla forces and other groups are securing supplies; weapons from Lebanon's civil war of the seventies and eighties have been shipped to Bosnia; surplus arms from Mozambique's civil war have been smuggled by former rebel soldiers to bands of criminals in Zimbabwe and South Africa. Many regions of the world find themselves awash in arms. Confronted with numerous unresolved social, economic, environmental, and political challenges, they are in effect sitting on a huge powder keg.[12]

★ ★ ★ ★

Since the end of the cold war, a series of arms control treaties have thinned out the overkill arsenals, but left substantial quantities of arms in place. Yet developing countries are not party to these agreements and not meant to be covered by them. The accords were tailored to the needs and circumstances of cold war protagonists in North America, Russia, and Europe; they have little relevance for the peculiar set of post–cold war conflicts and security issues.

First, whereas the East-West arms control treaties focus on major weapons systems, it is small arms and light weapons that are the principal killing tools of conflicts in the developing world. The trade and possession of such arms remain unregulated and unmonitored. Second, East-West–style arms control presupposes a stable relationship between opposing governments. Although the United States and the Soviet Union were highly antagonistic, there was considerable confidence that agreements hammered out by arms negotiators would indeed be carried out. In contrast, most contemporary conflicts involve a tangle of forces with conflicting claims of authority and divergent levels of accountability.[13]

Hence, the issue is not so much finding some magical balance-of-power formula to arrive at "stable" levels of armaments fielded by opposing sides as it is sharply curtailing the general availability of arms. The challenge is multiple: reducing the quantities of arms already in circulation, preventing a large-scale sell-off of surplus weapons, and restricting the production and transfer of new arms. This is only in part a task related to the weapons per se. To succeed, disarmament measures need to be embedded in broader social and economic policies that provide civilian livelihoods for demobilized soldiers and offer economic alternatives to companies, work forces, and communities dependent on arms production.

Without adequate programs to convert weapons production facilities to civilian use and help defense-dependent workers and communities make the difficult transition to an alternative livelihood, those affected will see little choice but to embrace the export option, even if the economics of it is as questionable as the politics. On the whole, conversion policies are rare. Governments are forced to prune their military industries, but are intent on protecting their defense industrial base. The emphasis is primarily on "dual use"—developing technologies with both civilian and military applications—rather than a decisive shift away from military use. The adjustment to lower military spending has largely been left to the market—resulting in corporate mergers, plant closures, and job loss. Although some companies are attempting to shift to civilian production, and local and regional authorities are trying to aid them and to help in the adjustment process, conversion as a deliberate tool of a disarmament policy is virtually nonexistent. And funds for this purpose are limited. This has created a vocal constituency for military projects of questionable merit.[14]

To date, efforts to collect and destroy surplus arms in countries emerging from violent conflict are highly inadequate. Yet they are needed in order to prevent the stocks from being used to unhinge peace settlements or being shipped to new trouble spots. In many countries, the large number of small arms in circulation poses a severe danger to internal security and economic development. Disarmament of former combatants has nominally been part of several U.N. peacekeeping missions, but it remains a largely unfulfilled task. In El Salvador, for example, peacekeepers collected and destroyed some 40,000 small arms from government and FMLN rebel soldiers, but as many as 300,000 weapons are believed to remain in circulation. In Mozambique, an estimated 6 million Kalashnikovs are still available.[15]

Some countries have tried so-called gun buy-back schemes, under which ex-combatants can turn in arms voluntarily in return for monetary or in-kind compensation. Buy-back programs were established in Panama, Haiti, and Nicaragua, and are under consideration in El Salvador. The United Nations was asked to help collect illicit arms in Mali, and has sent an advisory mission to determine the precise nature of the problem. Similar missions to Burkina Faso, Chad, Mauritania, and Niger are expected. In 1995, the U.N. General Assembly called on the Secretary-General to report on the buildup of small arms arsenals and ways to prevent it, and the first worldwide survey of the spread of small arms among civilians is being done by the United Nations. But some regions, including parts of South Asia and the Horn of Africa, are so inundated with arms and so characterized by a pervasive gun culture that the prospects of disarmament are severely clouded.[16]

Any buy-back or collection scheme can only succeed within broader community development programs and efforts to address crime and social breakdown. Demobilized soldiers and fighters will be much less inclined to keep their weapons and resort to criminal activities if they are successfully reintegrated into society. The prospects depend on the degree to which ex-combatants' civilian skills can be honed and whether there are adequate opportunities for employment and self-employment. In other words, the communities in which they live need to be economically revitalized. Demobilization of tens of thousands of soldiers has been undertaken in several countries, yet their reintegration remains a tremendous challenge.[17]

★ ★ ★ ★

Although many regions of the world are already awash in arms, preventing or at least limiting additional flows of weapons is crucial. The international arms trade has long been virtually a taboo subject. A breakthrough was achieved in December 1991, however, when the U.N. General Assembly established a U.N. Register of Conventional Arms to create greater transparency on arms transfers. Although this had been discussed as far back as the mid-sixties, it took the searing experience of the 1990 Iraqi invasion of Kuwait—demonstrating the dangers of allowing any country to assemble a huge war machine through unrestrained arms imports—to convince many governments of the merit of a Register.[18]

Governments are being asked to submit voluntarily information concerning imports and exports of major conventional weapons systems in seven categories—combat aircraft, attack helicopters, warships, battle tanks, armored vehicles, large-caliber artillery, and missiles. The first annual report (with data for 1992) was published in October 1993. Participation in the Register is slowly increasing, to 89 states in the 1994 Register, about half of all U.N. member-states. Although some leading importers, particularly in the Middle East, did not participate, the top exporters did. The number of governments submitting additional background information not expressly requested by the Register is also slowly increasing: 29 states provided information on equipment holdings and 20 states included data on national production.[19]

There are several proposals to expand the Register's quantitative and qualitative coverage. This might include steps such as increasing the number of weapons categories, including small arms; providing information about the sophistication of transferred weapons instead of just their numbers; and adding data on national production and stockpiles. (Also, regional registers could be created to provide data that are more specific and relevant to the area.)[20]

The Register might eventually grow into a tool for restraints against the flow of arms. Currently, submissions to the Register are simple after-the-fact reports. Although many governments can be expected to oppose any requirement for prenotification of pending sales, such information might at least be encouraged. Participation in the Register is now entirely voluntary, but it is to be hoped that full participation will eventually grow into a norm of good "global citizenship" for states. Countries not adhering to this norm, once it becomes widely accepted, might find it much harder to obtain arms or receive other rewards of international cooperation.

Changes in the Register will not come quickly; a recent review conference rejected a proposed expansion of its coverage. But concerned political leaders and grassroots groups are not waiting for governments to act. Former Costa Rican President and Nobel Peace Prize recipient Oscar Arias has initiated a campaign for a global "code of conduct" on arms exports. Under this, weapons would not be made available to countries that fail to hold free elections, to promote civilian control of the armed forces, to respect human rights, or to participate in the U.N. Arms Register, or to countries that engage in armed aggression. With the support of several other Nobel Peace laureates, Arias plans to present the code to the U.N. General Assembly in the fall of 1996.[21]

Citizen groups have pushed similar initiatives on the national level. A coalition of 275 grassroots groups in the United States has waged a campaign for several years; it has found considerable resonance in the U.S. Congress, but not enough support to pass legislation. In Europe, a number of national governments, parliaments, individual parliamentarians, and grassroots groups have called for a code of conduct. The European Parliament endorsed the idea in September 1992 and again in February 1994. The European Council of Ministers and the Conference on Security and Co-operation in Europe (CSCE) separately agreed on sets of criteria to govern arms exports of member-states. Yet their members still differ widely on how the criteria should be interpreted and whether governments should be legally bound by them.[22]

Oscar Arias has embedded his call for an arms transfer code of conduct in a broader Year 2000 Campaign to enhance the international community's capacity for demilitarization and conflict prevention. This also envisions special U.N. envoys to facilitate regional disarmament talks; substantial cuts in military expenditures and armed forces to be initiated by 2000; the use of the resulting savings for arms conversion, landmine clearance, community reconstruction, and the reintegration of demobilized soldiers; and debt forgiveness by industrial countries in exchange for demilitarization.[23]

To create confidence that far-reaching disarmament enhances rather than undermines security, effective inspection and monitoring arrangements are needed. The countries that belonged to NATO and the former Warsaw Pact now possess pools of experienced personnel and a range of well-developed verification technologies. But most other countries have no comparable capabilities. It would be useful to establish an international disarmament verification agency, in which expertise, resources, and responsibility for scrutinizing disarmament endeavors could be shared among the nations of the world.

★ ★ ★ ★

In sharp contrast to traditional security policies, strategies in the post–cold war era need to include conflict prevention in relations both between and within states, and hence a capacity to identify conflicts and potential crises before they erupt into large-scale violence. Through its peacekeeping operations and other projects, the United Nations is already deeply involved in efforts to create favorable conditions for the maintenance of peace. These include supervision of or advice on national elections, the establishment of representative and democratic national institutions, judicial reform, and the protection of human rights. Although many of these activities are being conducted in an effort to reestablish peace, they can be equally valuable in attempts to preserve it.[24]

But the United Nations does not have the capacity to issue early warnings about possible conflicts—nor does any other international organization, for that matter. Rising to that challenge would require a dedicated office to monitor and analyze particular events and underlying environmental, social, economic, and political trends that might signal impending large-scale conflict. One key job would be to prepare what former U.N. official Erskine Childers has called "boundary and ethnic contingency maps"—helping to identify potential "hot spots" where borders may be contested or contending groups may clash. Another task would be to develop more systematically some indicators of potential conflict, such as trends in soil erosion, water scarcity, unemployment, wage erosion, and human rights violations. (Some of these data are brought together each year in the *Human Development Report.*) But there also needs to be a mechanism that triggers preventive diplomacy once an early-warning alert has been issued to the Secretary-General and Security Council.[25]

Of course, signs of impending violent conflict are not always clearly discernible, and observers may legitimately disagree on the significance of particular developments. Early warning is not an exact science. It would be sensible for the United Nations to contract with outside experts, including university departments and nongovernmental organizations (NGOs) that know a great deal about particular countries and regions, for their insights and presence in the field. Also, existing U.N. field offices could be asked to provide relevant data to regional monitoring centers that in turn could relay information and findings to a central early-warning office. While interpreting the information presents a tremendous challenge, setting up an effective structure that gathers information should not present an insurmountable problem.[26]

In this context, it would be useful to establish standing impartial forums that routinely hear grievances, both within and among nations, and help to defuse tensions at an early stage, long before violent conflict is imminent. Regional conflict resolution committees and a professional corps of mediators could be set up by the U.N. Security Council in conjunction with regional organizations. In a similar vein, the Commission on Global Governance, an independent panel of distinguished leaders from around the world, proposed in 1995 that a Right of Petition be established at the United Nations for NGOs, and that a new Council for Petitions hear appeals that have a bearing on security issues and make recommendations to the U.N. General Assembly, Security Council, and Secretary-General.[27]

This conflict resolution infrastructure would be much stronger in the framework of regional organizations that promote confidence building, reductions in armaments, openness and consultation among neighboring countries, and respect for the rights of minorities. Though not nearly as strong as it might be, the Organization for Security and Co-operation in Europe (OSCE, the renamed CSCE) could serve as a model for this. The OSCE has set up a Conflict Prevention Center in Vienna and a Bureau for Democratic Institutions and Human Rights in Warsaw, has established a High Commissioner for National Minorities, and has dispatched a variety of long-term missions to conflict-torn member-states.[28]

Analysis of the difficulties with which the OSCE has to wrestle indicates the areas in which its effectiveness can be improved. First, the resources available to the organization are still quite limited. Second, questions remain about the pros and cons of its consensus-based decision-making process. Third, the national policies of the most powerful members, though nominally in accord with the OSCE, may be at variance with its own efforts. Despite these limitations, the OSCE is important as a forum for ongoing talks and negotiations, a tool to "civilianize" conflicts, and an instrument to strengthen international norms against the use of force.[29]

To date, other parts of the world lack effective regional mechanisms for conflict resolution. A group of prominent Africans, including former government leaders, proposed in 1991 the establishment of a permanent Conference on Security, Stability, Development, and Cooperation in Africa. But the Organization for African Unity, in summit meetings in 1993 and 1994, adopted only a watered-down version of the proposal, called a Mechanism for the Resolving of Conflicts. If conflict prevention and mediation are to take root as serious alternatives to warfare and pervasive insecurity, the capacities of regional organizations will need to be built up.[30]

To be sure, there are efforts—and successes—in mediating disputes or peacefully terminating violent conflicts. But typically these receive scant attention compared with the intense media scrutiny when things go wrong. And the attempts that are made are conducted in an ad hoc fashion. Success, where it comes, is achieved against all odds. In an alternative security system, such efforts would not be left to chance, or to the surprise initiative of individuals (such as former U.S. President Jimmy Carter's well-publicized efforts in Haiti, North Korea, Bosnia, and Central Africa). And they would focus on early involvement.

Although governments have on the whole done relatively little to promote conflict prevention systematically—either through the United Nations or other channels—NGOs have taken a lead in pursuing the issue, both in terms of research and practical, on-the-spot activities. The Carter Center's Conflict Resolution Program, based in Atlanta, Georgia, has an impressive track record. The Carnegie Corporation of New York set up the Carnegie Commission on Preventing Deadly Conflict in 1994, a study and discussion group composed of 16 international leaders and scholars. The Commission is working to identify both long-term, structural approaches (that is, policies that promote democratization, economic reform, education, and cross-cultural communication) and more immediate, practical steps to "build a firebreak against the outbreak and spread of mass violence." The Institute for Resource and Security Studies in Cambridge, Massachusetts, initiated a research and education project on Preventive Diplomacy and National Security. And International Alert, a British-based organization, is actively engaged in worldwide efforts to foster conflict mediation and preventive diplomacy.[31]

A growing number of organizations—governmental and non-governmental, national and international—are getting involved in promoting more democratic forms of governance and facilitating the emergence of more pluralistic, accountable societies. Actors as diverse as the U.S. National Endowment for Democracy, the Soros Foundation's Open Society Institute, the New York–based Committee to Protect Journalists, the German Friedrich Ebert Foundation, the new Stockholm-based Institute for Democracy and Electoral Assistance, and the U.N.'s Electoral Assistance Division are involved in a broad variety of activities. These range from political institution building and electoral support, to legislative and judicial assistance, to aiding the emergence of an active, viable civil society. No doubt the independence and effectiveness of these and other organizations varies. Yet many of these endeavors are likely to make valuable contributions to improving governance.[32]

Mediation and dispute resolution are key, but a more effective system must also be created to dissuade potential aggressors from carrying out acts of war and terror. For instance, if there had been a standing U.N. peacekeeping force to deploy along threatened borders, Saddam Hussein might have thought twice about invading Kuwait in 1990. Similarly, if there had been a force ready to protect civilians in "safe areas," Bosnian Serb forces would not have assaulted Bosnian civilians with impunity, and Hutu extremists in Rwanda would not have been able to engage in genocide.

Currently, U.N. peace operations are conducted entirely in an ad hoc fashion and hence are unable to engage in preventive actions. There is a burning need to establish a more permanent, reliable, rapidly deployable force. The United Nations should be empowered to recruit personnel for a specially trained standing unit of peacekeepers. These people would function not as combat personnel but as observers and mediators, and their deployment would show the international community's interest and active involvement in an emerging conflict. This permanent group could be backed up by national military units that remain under the jurisdiction of their respective governments, but that would be available to the United Nations on short notice. Trained and equipped so that they could easily interact with other national units placed on U.N. standby, they would be deployed along the border of a nation threatened by attack from a neighbor, or be used to protect civilian safe havens.[33]

There is an urgent need to work out generally applicable rules for when U.N. peacekeepers should intervene. Such rules would offer some assurance that intervention is dictated not by narrow political or economic interests of the more powerful governments but by a more universal concern for human dignity and well-being. And they would increase the likelihood that the international community acts in a timely fashion to prevent attacks against defenseless civilians even if the strategic interests of the big powers are not affected, instead of standing by, as happened in Rwanda.[34]

★ ★ ★ ★

Many of the elements of an effective peace system—demilitarization, conversion, peacekeeping, and conflict mediation—do exist in some form. But they are at no more than an embryonic stage, with rather uncertain prospects for their further development. Jonathan Dean, a former high-ranking U.S. diplomat who is now with the Union of Concerned Scientists in Washington, D.C., has captured the essence of an effective peace system: "The objective should be an international system that can increasingly do more—so that, in the long run, it would be called on to do less."[35]

9

A Human Security Budget

The main issue for security policy, Chapter 7 and 8 argue, is not how to produce deadlier weapons and advance military power in new ways, but how to reduce the availability of arms and alleviate the environmental, economic, and social stresses that generate insecurity. How are we to finance these endeavors?

Policies to prevent social breakdown, environmental degradation, and violent conflict require some substantial upfront investments, but they would cost much less than current reactive security policies. And they would help counter a trend toward ever rising expenditures on humanitarian emergencies—refugee assistance, relief supplies to civilian populations caught in war, food supplies to people victimized by drought and environmental decline, and so on.

Growing emergency spending—fighting the symptoms—is crowding out the kind of environmental improvement, social development, and peace investments that would help guarantee lasting human security, which would be curing the disease. As Jessica Mathews of the Council on Foreign Relations lamented in 1994, the world is "spending steadily more on the consequences of failed development and less on positive investment. Every year, more money is spent on disaster relief, humanitarian assistance, refugees and peace-keeping: on keeping the awful from becoming worse. Less and less is left for making the bad lastingly better."[1]

A look at the humanitarian expenditures of U.N. and other agencies confirms this grim assessment. The annual expenditures of the U.N. High Commissioner for Refugees have soared more than hundredfold since the early seventies, to about $1.3 billion in 1995. The office's budget is now the largest and fastest growing of the U.N.'s voluntarily financed programs. At the World Food Programme, meanwhile, relief activities for refugees and displaced persons climbed rapidly in the early nineties to claim two thirds of the budget; the agency's resources available for development shrank accordingly.[2]

Similarly, the share of UNICEF's budget devoted to emergencies climbed from 4 percent in 1990 to 28 percent in 1993 (and was only slightly lower in 1994). Humanitarian and emergency aid provided by members of the Organisation for Economic Co-operation and Development climbed more than tenfold—from about $300 million per year in the early eighties to $3.2 billion in 1994; relative to these countries' total development aid, the share of emergency spending climbed from less than 3 percent to about 10 percent during the same period. Similar trends are found in the expenditures of groups such as the Red Cross.[3]

With environmental issues, too, it is much cheaper—and preferable—to invest in programs to prevent degradation than to cope with its consequences. The costs of living with (rather than averting) climate change could be disastrously high; indeed, they would surely surpass the means of the poorest societies. Building defensive dikes to guard against sea level rise, resettling people from areas threatened by inundation, dealing with the agricultural repercussions of climate change, and confronting the devastation wrought by more frequent hurricanes and floods will impose tremendous costs.

A great deal of today's "peace operations" spending can also be characterized as emergency spending—belated efforts to douse the flames of conflict and alleviate its worst consequences. Peace-related expenditures have risen sixfold since the end of the cold war, to about $16 billion in 1994. But much of this goes to coping with the legacy of past and ongoing conflicts—helping war-torn societies get back on their feet, disposing of surplus weapons, clearing landmines, and conducting peacekeeping missions. By contrast, few resources are devoted to building institutions capable of enforcing and verifying restrictions on arms production and shipments; establishing forums that provide early conflict warning, peaceful settlement of disputes, and reconciliation services; and strengthening efforts to safeguard human rights and improve governance.[4]

If the international community were to move to an effective conflict prevention system, as mapped out in Chapter 8, it would likely reap substantial dividends over the years by avoiding recurring and rising expenditures for emergency-type peacekeeping and humanitarian aid.

★ ★ ★ ★

Figuring out how much it would cost to help bring about a future less reliant on military prowess obviously rests on a host of factors. Depending on the assumptions and scenarios, projections can vary substantially. The estimates presented here are in part based on actual experience. (See Table 9–1.) For instance, international landmine clearance, which is essential to permit societies to function again following civil conflict, is now funded to the tune of $70 million annually on the international level (plus some limited national funds). But experience suggests that outlays need to be boosted at least twentyfold. Hence, at a minimum, funding would need to reach about $1.5 billion a year.[5]

For other items, rough, order-of-magnitude estimates are presented. For a permanent, noncombat U.N. peacekeeping force of 10,000 persons, for example, personnel costs might come to $500 million a year. Assuming this force were deployed on 10 bases in each major region of the world, base operations and maintenance might reach another $500 million (base acquisition costs are put at zero, under the assumption that governments would make existing, surplus facilities available to the U.N. at no cost). Extrapolating from figures in a study by the Henry Stimson Center, training expenses could run to some $200 million annually. Equipment and transportation needs might add $1 billion. Finally, a central headquarters to direct the peacekeeping force could require $100 million more. All in all, this adds up to $2.3 billion per year.[6]

As indicated in Chapter 8, such a permanent, noncombat force would need to be backed up by standby forces. The Project on Defense Alternatives in Cambridge, Massachusetts, estimated that maintaining a U.N. military force that could continuously deploy a force of

Table 9–1. Estimated Annual Demilitarization and Peacebuilding Costs Worldwide

Institutions and Activities	Annual Cost (billion dollars)
Permanent U.N. peacekeeping force of 10,000 (incl. training, bases, equipment, operations center)	2.3
Backup peacekeeping forces (national units, on-call)	10
Conflict resolution and mediation centers[1]	1
Conflict early-warning network and analysis center[1]	1
Disarmament verification agency	1
Satellite monitoring agency (for early warning and disarmament verification purposes)	1
Dismantlement/disposal of surplus weapons	5 +
Landmine clearance	1.5 +
Arms conversion/adjustment	10
Demobilization/reintegration of soldiers	1
Post-conflict reconstruction and rehabilitation	5
Refugee repatriation	1
International criminal court	0.5
Human rights monitoring	0.25
Election observance/assistance	0.5
Total[2]	40

[1]Encompassing regional offices and a global headquarters.
[2]Rounded.
SOURCE: See endnote 5.

15,000 would require about 45,000 persons, when logistics and support personnel are included. (The Project envisions a standing U.N. army, in contrast to the proposal advanced here: forces that remain under their national governments' jurisdiction until called up.) This type of force would cost about $3.5 billion per year, including field operations. Using simple extrapolation from the Project's figures, a 50,000-strong standby force therefore might cost about $10 billion.[7]

These and the other figures in Table 9–1 are intended for broad illustrative purposes. What is most important is that these alternative security endeavors are eminently affordable—and at only a fraction of what national governments now spend on their military apparatuses. In fact, the total of $40 billion is only a little more than half the amount of money that governments spend each year just to figure out more sophisticated ways of killing enemies.[8]

By the same token, meeting social and environmental needs will not impose undue burdens in a world economy that now generates about $20 trillion in output each year. (See Table 9–2.) For example, estimates of implementing the two conventions agreed on at the Earth Summit in Rio—on climate change and biodiversity—range from $30 billion to $70 billion per year.[9]

Again, it is critical to understand that these expenditures are important investments that will yield benefits for decades to come. Compared with the routine worldwide expenditures for fossil fuel, it is clear that the costs of averting climate change by promoting energy efficiency and nonfossil sources of energy are far from prohibitive. Every year, the world spends some $450 billion on crude oil, $150 billion on natural gas, and $80 billion on coal. Including refining, processing, and distribution costs, the global fossil fuel budget may well equal $1 trillion annually.[10]

★ ★ ★ ★

Where would the money come from to pay for programs to enhance the social and environmental foundations of human security? We live in an era of anti-tax fervor and public belt-tightening, and hence the prospects for substantial additional resources are clouded at best. But al-

TABLE 9–2. *Estimated Annual Investments Required to Meet Urgent Social and Environmental Needs Worldwide*

Social/Environmental Need	Annual Cost (billion dollars)
Enhance energy efficiency	33
Prevent soil erosion	24
Provide adequate shelter	21
Eliminate malnourishment and starvation	19
Promote renewable energy sources	17
Stabilize population (family planning and reproductive health)	17
Provide safe, clean drinking water	15
Provide primary health care for all	15
Counter desertification	12
Manage forests sustainably	6
Preserve biodiversity	5
Cut illiteracy in half over 10 years	6
Enhance child survival, development, and protection	3
Phase out chlorofluorocarbons (to 2008)	1.8
Immunize children worldwide to prevent 1 million deaths annually	1.5
Total	196.3

SOURCE: Worldwatch Institute based on sources cited in endnote 9.

though some new money will be needed, large sums could be found by changing priorities within government budgets and shifting resources.

Military budgets are a case in point. Although global military expenditures have declined from their cold war peak—from about $1.25 trillion to perhaps about $800 billion (in 1993 dollars)—many governments still spend large sums of money and other resources building weapons that are no longer needed and maintaining armies poised against threats that have faded or disappeared. Even excluding Russia and China, the world's military producers still manufacture arms worth at least $150 billion each year. And governments worldwide pump an estimated $70 billion annually into inventing, developing, and designing more deadly weaponry. The alternative security policy outlined in Chapter 8 would permit substantial savings in military outlays and allow an increase in spending on demilitarization and improving social and environmental conditions.[11]

But such a shift in resources is far from automatic. Indeed, current funding for demilitarization is unreliable, and many developing countries have insufficient resources for the task at hand. Former Costa Rican President Oscar Arias has proposed the establishment of an International Disarmament Fund. He suggested earmarking each year a predetermined share of the money saved from reductions in the military budget of each nation—in effect capturing a portion of the peace dividend. These moneys could be enhanced by further contributing funds that are pegged to a certain percentage of military budgets. Table 9–3 presents, for illustrative purposes, how much such a formula could yield in annual revenues.[12]

★ ★ ★ ★

Concerning funding for sustainable development, a controversial debate has been going on around the issue of "additionality." Developing countries have insisted that environmental funds come on top of existing develop-

TABLE 9–3. *Hypothetical Contributions to a Global Demilitarization Fund, 1995-2000*[1]

	1995	1996	1997	1998	1999	2000	1995-2000
				(billion dollars)			
First-tier[2]	7.4	7.2	7.0	6.8	6.6	6.4	41.4
Second-tier[3]							
Industrial Countries	3.8	7.6	11.2	14.8	18.2	21.6	77.2
Developing Countries	0.4	0.7	1.1	1.4	1.7	2.0	7.3
Total	11.6	15.5	19.3	23.0	26.5	30.0	125.9

[1]Assuming military budgets will decline 3 percent a year. [2]1 percent of global military expenditures. [3]Set at 20 percent of savings in military expenditures for industrial countries and 10 percent for developing countries, relative to 1994 base year (military expenditures of industrial countries at $649 billion, of developing countries at $118 billion, and of world at $767 billion).
SOURCE: See endnote 12.

ment aid, rather than simply being reprogrammed—the equivalent of old wine in new bottles. But western countries have resisted the calls for more money.[13]

The experience with funding for international environmental endeavors, such as the Desertification Trust Fund, is that adequate money has not been forthcoming. The amount of new money available since the Rio conference has been trivial. There are currently two primary international funds for environmental issues that cannot be addressed exclusively in national terms. One is the Multilateral Ozone Fund, created in 1990 to help pay for the development of substitutes for ozone-depleting chemicals and the acquisition of equipment to recycle chlorofluorocarbons. Its establishment was critical to persuade countries like China and India to sign on to the Montreal Protocol on the phaseout of ozone-depleting substances. Donor governments pledged $240 million for 1991–93 and an additional $455 million for 1994–96. Yet actual payments have fallen short of the commitments, and the pledges themselves may be insufficient. Russia alone, for instance, will need some $600 million over four years to meet the provisions of international ozone accords.[14]

The second fund is the Global Environment Facility (GEF), set up in 1991 as a $1.3-billion pilot project to fund investments on preserving the global commons and, later, to finance implementation of the climate change and biodiversity treaties. GEF became a permanent body in 1994, with an additional $2 billion pledged for the period up to 1997. Although GEF has an independent secretariat, it is managed jointly by the U.N. Development Programme (UNDP), the U.N. Environment Programme, and the World Bank. The decision-making arrangements for the governing council are intended to make it responsive to both donors and recipients. How well GEF will work remains to be seen. It is clear that the amounts of money involved so far are trivial compared with estimates of resources needed, and funding even at these limited levels is far from assured. GEF's real significance may lie elsewhere. It could play the role of catalyst for mobilizing and leveraging additional resources in both public and private sectors, including those of the World Bank.[15]

★ ★ ★ ★

Development aid can play an important role in improving human security, to the extent that it helps to alleviate poverty, overcome social inequities, and build indigenous capacities. Yet there are problems with both the quantity and the quality of aid. Too often, aid money flows into questionable "white elephant" projects, is tied to purchases from donor countries, or is used to curry influence with a strategic ally. The neediest are often not the beneficiaries: according to the *Human Development Report 1994,* two thirds of the world's 1.3 billion poor live in 10 countries that together receive less than one third of official development assistance (ODA). Governments spending excessively on their military are reaping the greatest rewards: in 1992, countries that spent more than 4 percent of their gross national product (GNP) on the military received $83 of aid per capita compared with $32 for countries that spent less than 2 percent. Data compiled by the World Development Movement in London show that since the mid-eighties, U.K. aid to countries that are not major buyers of British arms, such as Tanzania, has fallen or stagnated, while it has grown for countries such as Indonesia that are big customers.[16]

Official development assistance from most Western countries has persistently fallen short of an international goal set in the seventies and reaffirmed at the World Social Summit in Copenhagen in 1995: to disburse the equivalent of 0.7 percent of donor countries' GNP. In fact, by 1993 ODA had declined to the equivalent of 0.29 percent, the lowest figure since 1973. And proponents of foreign aid now have to reckon increasingly with "donor fatigue." In the United States, ideological opposition to foreign aid is growing. Had the 0.7 percent goal been reached, development aid in recent years would have amounted to about $130 billion a year instead of the actual level of about $55–65 billion.[17]

As in the case of resources for environmental protection, donors are resisting "additionality." Even though the official documents of the World Social Summit incorporate calls for "new and additional resources," for instance, developing countries' proposal to create an International Fund for Social Development was omitted.[18]

At least as important as the total amounts of money available, however, is the issue of what development aid is actually designed to accomplish. As the *Human Development Report 1994* argues, aid is often not targeted at priority areas of human development (poverty reduction, basic education, primary health care, rural water supplies, nutrition programs, and family planning services). UNDP has proposed a 20:20 Compact on Human Development to help focus a larger portion of existing aid flows on these priority areas. Under the Compact, developing countries are asked to raise the share of their national budgets devoted to these concerns from 13 to 20 percent, and donor countries are asked to increase the share of their aid budgets allocated to such programs from 7 to 20 percent. This would increase the money going to human priority areas by about $40 billion a year. The 20:20 Compact received a qualified endorsement at the World Social Summit, but it remains to be seen how sincere national governments will be when they translate their rhetoric into reality.[19]

★ ★ ★ ★

Debt relief, as discussed in Chapter 7, could be another mechanism for mobilizing human security funds. Developing countries, especially the poorest and most-indebted ones, would be better able to marshal resources for human security purposes if their foreign debt burden were substantially lighter. From 1987 to 1993, developing countries paid $1.3 trillion to service their debts, yet their total outstanding debt grew by one third. There has been a very large amount of debt rescheduling and refinancing (adding up to some $509 billion during the same period). But a comparatively trivial $30 billion of outstanding principal was forgiven, mainly for loans whose repayment was under any circumstances extremely doubtful. Nongovernmental groups have pioneered a variety of innovative debt reduction efforts, such as debt-for-development and debt-for-nature swaps, but their scope relative to the overall size of foreign debt has remained marginal. Developing countries' foreign debt has been reduced by less than $1 billion through such swaps.[20]

Without a more determined debt relief strategy, the hemorrhaging of many developing countries' economies will continue and their human insecurity will keep growing. Particularly for the 32 severely indebted low-income countries, as classified by the World Bank, debt write-off could be an important contribution to their stabilization. Their debt service runs to $7–10 billion a year, about 20–25 percent of their export revenues. In 1987–93, $49 billion worth of their debt was rescheduled; only $6 billion was forgiven. Most of the 32 countries are in sub-Saharan Africa, which is now paying its external creditors more each year than it spends on health and education programs combined. Tanzania, for example, paid $155 million in 1995 to service its debt, more than twice what the government spent to provide clean water to its citizens.[21]

It is time for lenders to cancel the bulk of the debts owed by the 32 countries—some $200 billion, or about 10 percent of all developing countries' foreign debt. They might do so in exchange for a "human security conditionality": commitments by the debtors to reduce their military expenditures and armed forces, and to invest in social and environmental areas resources that otherwise would have gone into debt service or to the armed forces.[22]

★ ★ ★ ★

Financial contributions by national governments to international funds and organizations are notoriously unreliable. This is true even for payments that are mandatory under international law, such as the membership assessments for the U.N. budget. Proposals for global funds for demilitarization, environmental protection, and social development have been given short shrift. Given this reality, and the unpleasant fact that funding for global endeavors is extremely vulnerable to the political pressures of the day, many observers have called for alternative sources of revenue for the United Nations and a variety of other international programs.

Among the possible sources of funding are fees levied on several international activities, including air travel, maritime shipping, telecommunications, and trade (including arms sales). There is no shortage of ideas, but rather a paucity of action. Norwegian Prime Minister Gro Harlem Brundtland stated at the Social Summit: "We need reliable sources of finance, assessed contributions, even exotic new sources of finance. We are not poor in ideas but close to bankrupt in their implementation."[23]

A prominent proposal is the so-called Tobin tax. Put forth by 1981 Nobel Economic Prize winner James Tobin, this would levy a tax on foreign-exchange transactions to dampen rapidly growing speculative capital movements and to control market volatility. Since international capital transactions rely on having a workable system of international governance, the Tobin tax would in effect constitute a service fee. As much as $1–1.5 trillion denominated in the world's major currencies changes hands every day, so even a very small levy would yield substantial revenues: a tax of 0.05 percent on the value of each transaction would yield $180–270 billion per year.

How such revenues are used would still be subject to governments' collective decision making, yet the Tobin tax would provide a reliable source of funding and remove the possibility that individual nations could withhold payments as a weapon. The Tobin tax, or some variant of it, is essential for effective global governance and human security, as the independent Commission on Global Governance noted in 1995.[24]

As the Tobin tax implicitly acknowledges, public flows of money are only part of the picture. When confronted with developing countries' demands for additional resources, western governments often point to the potential of private funding: at some $167 billion in 1995, private foreign direct investment was about three times as large as ODA, a ratio likely to grow even larger in coming years.[25]

But merely pointing out this fact amounts to an abdication of responsibility. Public policy needs to set the parameters within which private flows of money can serve environmental and social needs far better than they do now. National legislation can set rules and provide incentives to channel private capital into energy efficiency or the development of solar and wind power technologies, for example, and to promote job creation. Governmental export-import financing and foreign investment risk insurance programs likewise need to incorporate environmental and social criteria more centrally. And international codes of conduct need to set appropriate norms and standards for both private investment and private bank lending. Since private companies are driven by the profit motive, rather than by a social or environmental conscience, it is up to governments—both individually and in concert with each other—to establish proper rules and regulations.

★ ★ ★ ★

The wrangling between industrial and developing countries over additional resources and the proper role of public versus private flows of money should not be allowed to mask an essential truth. Put simply, the choice before humanity is pay now or pay much more later. The cost of failing to advance human security is already escalating. Devoting greater resources to demilitarization, environmental sustainability, and social well-being may be regarded as incurring unwelcome expenses. Yet they constitute a set of highly beneficial, mutually reinforcing, and long overdue investments. Experience already suggests that failing to make these investments will result only in greater costs—and massive suffering—later on.

It is time to strike a new balance in our security investments: a balance that allows the international community to curtail the excessive reliance on traditional military means, correct massive social and environmental investment deficits, and avoid costly emergency measures.

10

A Global Partnership for Human Security?

In his widely read 1994 *Atlantic Monthly* article, "The Coming Anarchy," journalist Robert Kaplan uses the image of a stretch limousine rolling along potholed streets to describe what is happening worldwide today. Travelling in this luxurious, air-conditioned vehicle, the well-off bypass desolate zones inhabited by the downtrodden. It is an apt metaphor: the richer classes and societies are tempted to pretend that they can simply roll up the limo's tinted windows and ignore the squalor and insecurity, the problems and challenges, that surround them.[1]

"Rolling up the windows" takes a number of forms. It can be found in walled-in, guarded compounds for the wealthy that are off-limits to nonresidents in countries as diverse as the United States and Brazil, in the provision of drinking water and electricity through private wells and generators to elites in Pakistan while the population at large depends on public systems that are falling apart, and in the efforts of wealthy countries to erect walls or fences at their borders to keep out unwanted migrants and refugees.[2]

Looking through the rose-colored windows of the stretch limo may provide temporary comfort to the well-off, but it is largely an illusion. Despite enormous disparities, the world is increasingly becoming one. As documented throughout *Fighting for Survival*, in many crucial respects communities and entire nations are no longer the sole masters of their destinies. Production, trade, investment, modern communications, and travel are now inherently global in scale, rapidly transforming this diverse planet into an interlinked unit. More and more often, events in one corner of the globe either have immediate repercussions elsewhere or eventually filter through to other regions. And though the signs of social and environmental stress are much more pronounced in some locations, all countries face common challenges.

The nations of the earth are inextricably linked by the fate of the biosphere. No matter how wealthy and technologically sophisticated, no country can escape the effects of climate change or ozone depletion; it does not matter whether a ton of carbon or chlorofluorocarbons was emitted in China or Brazil, or in the United States or Japan. No country can fend off air- or waterborne contaminants generated by neighbors upwind or upstream. None can erase or undo the consequences of dwindling biodiversity. As Soviet Foreign Minister Eduard Shevardnaze rightly said in the late eighties, "the biosphere recognizes no division into blocs, alliances, or systems. All share the same climatic system and no one is in a position to build his own isolated and independent line of defense."[3]

Richer, industrial countries depend less immediately on the integrity of natural systems for jobs and income; they have much more of a buffer than countries whose populations overwhelmingly depend on agriculture and resource extraction for their livelihoods. A more diversified, vibrant economy, technological prowess, and financial wherewithal permit richer countries to cope better with environmental and social pressures and to devise ways to circumvent or delay scarcities. But this may convey a false sense of security. In the final analysis, rich and poor alike depend on the stability of climate and weather patterns, the fertility of agricultural lands, the availability of water for irrigation, and the resilience of ecosystems.

Nor can the better-off ignore the destabilizing effects of massive environmental degradation, often intertwined with profound social and economic disparities and unresolved or reemerging ethnic tensions. Increasingly, the consequence is a violent unraveling of entire societies that unleashes large numbers of refugees, turns immense swaths of territory into emergency zones, and triggers shock waves of distress in adjacent countries.

The effects on others, near and far, are many. First, whether they relish the fact or not, other countries wind up hosting large numbers of refugees, possibly for many years. Second, they are drawn into extensive humanitarian relief operations that are difficult and time-consuming, and that divert attention and energy from other tasks. Third, they frequently have little choice but to engage in usually frustrating efforts to negotiate fragile cease-fires, deploy peacekeepers, and reestablish at least a semblance of order. Finally, all these consequences have substantial and rapidly rising financial costs. Emergency expenditures eat up resources otherwise available for social and economic advancement and environmental protection.

But even short of cataclysmic events, industrial countries are likely to experience several "boomerang" effects from unalleviated social, economic, and environmental pressures in developing countries.

Nations that are chronically unstable or in the grip of social and environmental convulsions are in effect "lost markets" to traders and investors. Instability and conflict threaten a serious disruption of economic activity and loss of investment. Industrial countries, with extensive commercial and financial links virtually everywhere, have an inherent interest in working to avoid conflicts that threaten to disrupt these links. U.S. Secretary of State Warren Christopher recognized this in a 1996 memorandum to his Under and Assistant Secretaries, stating that environmental decay "makes for weaker trading partners and for greater reliance on foreign assistance."[4]

Because farmers in countries like Bolivia, Peru, and Colombia have great difficulty securing access to markets for their food crops or selling them at an adequate price, for example, they often have little choice but to grow coca and other illegal drugs. These crops are clearly lucrative: one hectare of coffee in Bolivia brings in $46 in export revenues for farmers, but coca yields $133. Bolivian coca production has risen along with poverty as the country has undergone severe "structural adjustment," and as credits for food crops produced by small peasants have dried up. An estimated 20 percent of Bolivia's national income is now accounted for by illicit coca exports.[5]

The boomerang occurs when the United States and other industrial countries become the main destinations of the $500-billion-a-year international drug trade. The flow of drugs into these countries is now increasingly seen as a "national security" issue, with all the attendant temptations to militarize any attempted solution.[6]

As developing countries become more tightly integrated into the global economy, the expectation is that they will quickly shed the characteristics of underdevelopment. Nowhere has this expectation been more nurtured—and then dashed—than in Mexico. Yet Mexico remains a country of fundamental economic weaknesses and enormous social inequities. The illusions that it would quickly join the ranks of older industrial countries were shattered by the severe peso crisis of 1994/95. The potential repercussions for the international financial system were such that the United States had little choice but to come to the rescue with a $50-billion bailout. In return, Mexico was "effectively giving Washington veto power over much of Mexico's economic policy for the next decade," as *New York Times* correspondent David Sanger wrote—surrendering a key aspect of its national sovereignty.[7]

The severe contraction of the Mexican economy that came with the peso crisis dealt a heavy blow to ordinary Mexicans, further fueling the incentives to migrate—primarily to the United States—in search of a better livelihood. The influx of Mexicans, in turn, is being exploited by U.S. politicians who think they can make political capital out of fanning anti-foreigner sentiments. Unless dealt with more forthrightly, economic roller coasters, as well as the forces of war, repression, and environmental degradation, will continue to uproot a large and growing number of people and confront richer societies.

Another boomerang effect is found in the legacy of arms proliferation. Industrial countries in both East and West have long sold arms with little regard to the consequences. The assumption was, and largely still is, that pumping large amounts of arms into areas such as Afghanistan, the Persian Gulf, the Horn of Africa, or Central America would not affect the countries doing the selling. Yet many of these arms have disappeared into the black market, only to turn up as a destabilizing factor in other regions. And it is no longer just developing countries that receive illicit arms transfer. In May 1996, representatives of two Chinese state-owned arms manufacturing companies were arrested in the United States on charges of smuggling 2,000 AK-47 automatic rifles into the country. These and planned additional shipments of larger-caliber arms were apparently intended to supply "American gangs in need of heavy weapons to wipe out their rivals," according to the *New York Times*.[8]

Concern is also growing that sufficient amounts of weapons-grade fissile materials to assemble at least a small nuclear device might find their way to "rogue" states or to rebel groups that are prepared to use it or at least threaten to do so. These concerns are nourished by the continued instability in successor states of the Soviet Union, the erosion of control over fissile materials, and the temptations to smuggle materials among weapons scientists and technicians who now face an uncertain economic future.

What these very diverse examples show is that a common, global approach to social, economic, environmental, and military issues is essential. Yet convincing governments to form a global partnership for human security—and to initiate appropriate national policies to that end—remains an uphill struggle. Parochial, xenophobic posturing brings short-term political gain and helps cloud the longer-term implications. When contemporary challenges are seen as the problems of "others," there is a strong temptation among "the haves" to stick to containment and damage control. Western reaction to the phenomenon of ethnic conflicts in developing countries is symptomatic: these confrontations are now frequently portrayed as the outgrowth of irrational "ancient hatreds" without remedies, which are best left to be fought out by the protagonists. This attitude ignores the role that underlying social, economic, and environmental factors have played in the outbreak of hostilities, not to mention that of outside arms supplies.[9]

In "The Coming Anarchy," Robert Kaplan lamented: "Mention 'the environment' or 'diminishing natural resources' in foreign-policy circles and you meet a brick wall of skepticism or boredom." Nearly three years later, the brick wall is showing signs of cracks, but it is far from crumbling.[10]

Both nationally and internationally, some progress toward a human security policy is being made. But the pace is slow and uncertain, as demonstrated by the Clinton administration's effort to inject environmental criteria more prominently into the foreign policy agenda. Soon after coming into office, the administration created the post of Under Secretary for Global Affairs within the State Department, a position filled by former Senator Timothy Wirth. Yet this and other initiatives met with considerable skepticism and opposition within the State Department, in Congress, and among national security analysts.[11]

Traditionalists sneer at environmental considerations as "soft" issues—as compared with the "hard" issues and means (weaponry and balance-of-power considerations) central to traditional foreign and security policy. Professor Angelo Codevilla of Boston University, a former Reagan administration official, contended in late 1995 that "the hard stuff is as important as ever because guns will determine the future of the world as much as they ever did."[12]

It was not until early 1996 that Secretary of State Christopher instructed all State Department bureaus and embassies to develop plans to incorporate environmental issues into global, regional, and bilateral U.S. diplomacy. Beginning in 1997, the State Department is to issue an annual report on Global Environmental Challenges. It is also working to improve interagency coordination with the Environmental Protection Agency and the Departments of Defense, Energy, Commerce, Interior, and Agriculture. It remains to be seen whether these efforts will amount to more than just window dressing.[13]

The struggle within the State Department over making environment a higher priority indicates just how difficult the reorientation of traditional foreign and security policies is. To use the imagery of old, the battle has been joined, but the war remains to be won. This is particularly the case since much greater change lies ahead if governments are to move beyond the realm of slogans and speeches. Adding environmental and other nontraditional factors to an already long list of criteria and objectives is one thing. But transforming national policy so that it is less guided by outdated perceptions of exclusive national interest and more by an understanding of the global human interest is quite another.

The transformation will require a far-reaching shake-up and reorganization of foreign and security policy bureaucracies, and a hardheaded assessment of the value and purpose of existing agencies and ministries. And it will demand that the extreme compartmentalization among ministries responsible for various facets of a human security policy—including defense, foreign affairs, foreign aid, finance, and energy—be addressed.

* * * *

The notion that growing interdependence generates a strong common interest in global cooperation on economic, environmental, and security issues is no longer new. The case has been made by, among others, a series of high-profile commissions composed of international leaders of exceptional stature, such as the 1980 Brandt Commission on North-South issues, the 1982 Palme Commission on disarmament and security, the 1987 Brundtland Commission on environment and development, and the 1995 Commission on Global Governance.[14]

Yet policy-making in the fields of social development, environment, and peace and security has a long way to go before it genuinely reflects the insights of these commissions. Interdependence increasingly seems to dictate a global community of interests. But because it has come about by default more than by design, it has not automatically generated the political will needed for greater international cooperation. The world is still plagued by enormous inequalities in power, wealth, and capacity to influence global affairs, just as it is marked by divided loyalties, conflicting interests, and fragmentation along political, ethnic, religious, and social dividing lines.

Nations have "outgrown" their borders, but many of the structures and strictures of nation-states remain firmly in place. Like membranes, national borders are at times porous, at times impermeable. They cannot stem the flow of goods, money, or ideas any more than they can that of polluted air or water. Still, they often block appeals for common approaches to shared problems. Visions of a common, human security system are still held in check by the reality of national rivalry. Thus, paradoxically, we live in a universe without firm boundaries, and a planet that retains many of its borders.

In principle, human security offers a more fruitful basis for cooperation and security among nations than traditional notions of military security because it is a positive and inclusive concept. While military security rests firmly on the competitive strength of individual countries at the direct expense of other nations, human security cannot be achieved unilaterally: it requires and nurtures more stable and cooperative relationships among nations, and depends on greater solidarity among classes and communities.

The task of strengthening the social, economic, and environmental dimensions of global security is as challenging as it is imperative. To succeed, no less than a fundamental reexamination of the assumptions that have long guided national security policies is required. The profound transformations that our world is undergoing continue to challenge the traditional conduct of diplomacy and the established forms of governance. Indeed, "national security" as such has become an outmoded concept: either security is attained through the difficult process of global cooperation, or it will remain elusive.

The thinking that has brought walled-in communities of the rich, private electricity and water supplies in the midst of impoverished communities, and attempts to seal borders against unwanted populations will not fade away overnight. It will take continued public education, grassroots empowerment, and inspired leadership to overcome the most fundamental barriers to change of all: those of the mind.

Notes

Chapter 1. The Transformation of Security

1. U.N. Development Programme (UNDP), *Human Development Report 1994* (New York: Oxford University Press, 1994).
2. Victoria Holt, *Briefing Book on Peacekeeping: The U.S. Role in United Nations Peace Operations* (Washington, D.C.: Council for a Livable World Education Fund, 2nd ed., 1995).
3. Woolsey cited in Theo Sommer, "Keiner Will den Weltgendarm Spielen," *Die Zeit*, June 17, 1994.
4. Figure 1–1 from Hal Kane, "Wars Reach a Plateau," in Lester R. Brown, Nicholas Lenssen, and Hal Kane, *Vital Signs 1995* (New York: W.W. Norton & Company, 1995).
5. Internal conflicts since 1945 from Aaron Karp, "Small Arms—The New Major Weapons," in Jeffrey Boutwell, Michael T. Klare, and Laura W. Reed, eds., *Lethal Commerce: The Global Trade in Small Arms and Light Weapons* (Cambridge, Mass.: American Academy of Arts and Sciences, 1995); share of civilian victims from Ruth Leger Sivard, *World Military and Social Expenditures 1989* (Washington, D.C.: World Priorities, 1989), and from Ernie Regehr, "A Pattern of War," *Ploughshares Monitor*, December 1991.

6. UNICEF, *The State of the World's Children 1996* (New York: Oxford University Press, 1996).
7. Dan Smith, *War, Peace and Third World Development*, Human Development Report Office, Occasional Papers No. 16, UNDP, New York, 1993; Michael T. Klare, "The Global Trade in Light Weapons and the International System in the Post-Cold War Era," in Boutwell, Klare, and Reed, op. cit. note 5; Christopher Louise, *The Social Impacts of Light Weapons Availability and Proliferation*, Discussion Paper No. 59, U.N. Research Institute for Social Development, Geneva, March 1995.
8. See, for example, Gerald B. Helman and Steven R. Ratner, "Saving Failed States," *Foreign Policy*, Winter 1992–93.
9. James N. Rosenau, "New Dimensions of Security: The Interaction of Globalizing and Localizing Dynamics," *Security Dialogue*, September 1994.
10. Samuel P. Huntington, "The Clash of Civilizations?" *Foreign Affairs*, Summer 1993. For a lucid critique, see Richard E. Rubenstein and Jarle Crocker, "Challenging Huntington," *Foreign Policy*, Fall 1994.
11. Figure of 40 percent from Rubenstein and Crocker, op. cit. note 10; half of all countries experienced ethnic strife from UNDP, op. cit. note 1; other data from Ted Robert Gurr, *Minorities at Risk: A Global View of Ethnopolitical Conflicts* (Washington, D.C.: U.S. Institute for Peace Press, 1993).
12. Erskine Childers, "UN Mechanisms and Capacities for Intervention," in Elizabeth G. Ferris, ed., *The Challenge to Intervene: A New Role for the United Nations?* Conference Report 2 (Uppsala, Sweden: Life and Peace Institute, 1992); Human Rights Watch, *Slaughter Among Neighbors: The Political Origins of Communal Violence* (New Haven, Conn.: Yale University Press, 1995).
13. Ted Robert Gurr, "Third World Minorities at Risk Since 1945," Background Paper on the Conference on Conflict Resolution in the Post–Cold War Third World, U.S. Institute of Peace, Washington, D.C., October 3–5, 1990.
14. Stephen D. Goose and Frank Smyth, "Arming Genocide in Rwanda," *Foreign Affairs*, September/October 1994.
15. Rosenau, op. cit. note 9.
16. Independent Commission on Disarmament and Security Issues, *Common Security* (London: Pan Books Ltd., 1982); Simon Dalby, "Ecopolitical Discourse: 'Environmental Security' and Political Geography," *Progress in Human Geography*, Vol. 16, No. 4, 1992; Geoffrey D. Dabelko and David D. Dabelko, "Environmental Security: Issues of Conflict and Redefinition," *Environmental Change*

and Security Project (Woodrow Wilson Center, Washington, D.C.), Spring 1995; UNDP, op. cit. note 1.

17. William Eckhardt, "War-Related Deaths Since 3000 BC," *Bulletin of Peace Proposals*, Vol. 22, No. 4, 1991.

CHAPTER 2. Environmental Stress

1. Stefan Klötzli, "The Water and Soil Crisis in Central Asia—A Source for Future Conflicts?" Occasional Paper No. 11, Environment and Conflicts Project (ENCOP), Bern, Switzerland, May 1994; Sandra Postel, "Where Have All the Rivers Gone?" *World Watch*, May/June 1995; "Aral Sea Fact Sheet," distributed at Dangerous Waters: Geography, Politics, and Environment in Critical Water Systems, Columbia University, School of International and Public Affairs, New York, May 22–23, 1995.
2. Eileen Claussen, "Environment and Security: The Challenge of Integration," in *Environmental Change and Security Project Report* (Woodrow Wilson Center, Washington, D.C.), Spring 1995; Astri Suhrke, "Pressure Points: Environmental Degradation, Migration and Conflict," in Occasional Paper No. 3, Project on Environmental Change and Acute Conflict, American Academy of Arts and Sciences and University of Toronto, March 1993; Norman Myers, *Ultimate Security: The Environmental Basis of Political Stability* (New York: W.W. Norton & Company, 1993).
3. Desertification is defined by UNEP as "land degradation in arid, semi-arid, and dry subhumid areas resulting mainly from adverse human impacts"; Günther Bächler, "Desertification and Conflict: The Marginalization of Poverty and of Environmental Conflicts," Occasional Paper No. 10, ENCOP, Bern, Switzerland, March 1994; percentage of degraded agricultural land from L.R. Oldeman, International Soil Reference and Information Centre, Wageningen, Netherlands, private communication with Gary Gardner, Worldwatch Institute, Washington, D.C., September 21, 1995.
4. Annual loss from Myers, op. cit. note 2; per hectare loss from Gary Gardner, "Preserving Agricultural Resources," in Lester R. Brown et al., *State of the World 1996* (New York: W.W. Norton & Company, 1996).
5. Table 2–1 is based on Bächler, op. cit. note 3; regional soil degradation from L.R. Oldeman, International Soil Reference and Information Centre, Wageningen, Netherlands, private communication with Gary Gardner, Worldwatch Institute, Washing-

ton, D.C., April 12, 1996; crisis areas from Suhrke, op. cit. note 2; Mexico from Myers, op. cit. note 2.

6. Interior soil loss from Vaclav Smil, "China's Environmental Refugees: Causes, Dimensions and Risks of an Emerging Problem," in Kurt R. Spillmann and Günther Bächler, eds., "Environmental Crisis: Regional Conflicts and Ways of Cooperation," Occasional Paper No. 14, ENCOP, Bern, Switzerland, September 1995; threats to China's food self-sufficiency from Lester R. Brown, *Who Will Feed China? Wake-Up Call for a Small Planet* (New York: W.W. Norton & Company, 1995).
7. Myers, op. cit. note 2; Claussen, op. cit. note 2.
8. Sandra Postel, *Last Oasis: Facing Water Scarcity* (New York: W.W. Norton & Company, 1992); Myers, op. cit. note 2; M. Abdul Hafiz and Nahid Islam, "Environmental Degradation and Intra/Interstate Conflicts in Bangladesh," Occasional Paper No. 6, ENCOP, Bern, Switzerland, May 1993.
9. Hafiz and Islam, op. cit. note 8; Michael Renner, *National Security: The Economic and Environmental Dimensions*, Worldwatch Paper 89 (Washington, D.C.: Worldwatch Institute, May 1989).
10. Suhrke, op. cit. note 2; Alan Thein Durning, *Guardians of the Land: Indigenous Peoples and the Health of the Earth*, Worldwatch Paper 112 (Washington, D.C.: Worldwatch Institute, December 1992).
11. Postel, op. cit. note 8; Table 2–2 is derived from Peter H. Gleick, "Water and Conflict," in Occasional Paper No. 1, Project on Environmental Change and Acute Conflict, American Academy of Arts and Sciences and University of Toronto, September 1992. As reflected in Table 2–2, water-scarce countries are mostly found in Africa and Asia (specifically the Middle East); there are few countries that belong in this category in the Americas, Oceania, and Europe (although Belgium, the Netherlands, Hungary, and Malta are included).
12. Postel, op. cit. note 8; Sandra Postel, "Forging a Sustainable Water Strategy," in Brown et al., op. cit. note 4.
13. Postel, op. cit. note 8.
14. Anne E. Platt, "Aquaculture Boosts Fish Catch," in Lester R. Brown, Nicholas Lenssen, and Hal Kane, *Vital Signs 1995* (New York: W.W. Norton & Company, 1995); U.N. Food and Agriculture Organization (FAO) from Peter Weber, *Net Loss: Fish, Jobs, and the Marine Environment*, Worldwatch Paper 120 (Washington, D.C.: Worldwatch Institute, July 1994).
15. FAO from James Harding and Deborah Hargreaves, "Fish Knives Out in Defence of Canada's Turbot," *Financial Times*, March 11/12, 1995; Newfoundland from Bernard Simon, "Tur-

bot-Charged Dispute," *Financial Times*, March 18–19, 1995; Weber, op. cit. note 14.

16. William J. Broad, "Creatures of the Deep Find their Way to the Table," *New York Times*, December 26, 1995.
17. "New Hope for Species Depleted by Overfishing," *New York Times*, August 27, 1995; Günter Bächler et al., *Umweltzerstörung: Krieg oder Kooperation?* (Münster, Germany: Agenda Verlag, 1993); World Resources Institute study from "Half of World's Coastlines Are Found to Be in Peril," *New York Times*, November 7, 1995.
18. Weber, op. cit. note 14.
19. "New Hope for Species Depleted by Overfishing," op. cit. note 17; Weber, op. cit. note 14.
20. Bächler et al., op. cit. note 17; Weber, op. cit. note 14; Deborah Cramer, "Troubled Waters," *Atlantic Monthly*, June 1995.
21. "Ozone Hole Grows Larger as Levels Drop 10%," *New York Times*, September 13, 1995; William K. Stevens, "Study of Cloud Patterns Points to Many Areas Exposed to Big Rises in Ultraviolet Radiation," *New York Times*, November 21, 1995.
22. "U.S. Global Change Research Program Second Monday Seminar Series: Anthropogenic Ozone Depletion—Status and Human Health Implications," Seminar Announcement, in the APC electronic conference igc:climate.news, November 2, 1995.
23. Myers, op. cit. note 2.
24. Anjali Acharya, "CFC Production Drop Continues," in Lester R. Brown, Christopher Flavin, and Hal Kane, *Vital Signs 1996* (New York: W.W. Norton & Company, 1996); "U.S. Global Change Research Program," op. cit. note 22; "New Data Point to the Ultimate Recovery of the Ozone Layer," *New York Times*, May 31, 1996.
25. William K. Stevens, "U.N. Warns Against Delay in Cutting Carbon Dioxide Emissions," *New York Times*, October 25, 1995; William K. Stevens, "Talk About Weather: U.N. Says People Do Something About It," *New York Times*, December 1, 1995.
26. Stevens, "U.N. Warns Against Delay," op. cit. note 25; Christopher Flavin, "Facing Up to the Risks of Climate Change," in Brown et al., op. cit. note 4.
27. Stevens, "U.N. Warns Against Delay," op. cit. note 25; number of people at risk from Chapter 9, "Coastal Zones and Small Islands," in Robert T. Watson, Marufu C. Zinyowera, and Richard H. Moss, eds., *Climate Change 1995: Impacts, Adaptations and Mitigation of Climate Change: Scientific-Technical Analyses* (Cambridge: Cambridge University Press, for the Intergovernmental Panel on Climate Change, 1996).

28. Hafiz and Islam, op. cit. note 8; Myers, op. cit. note 2; *char* from Suhrke, op. cit. note 2.
29. Myers, op. cit. note 2; William K. Stevens, "Scientists Say Earth's Warming Could Set Off Wide Disruptions," *New York Times*, September 18, 1995.
30. Myers, op. cit. note 2; rice production impacts from Watson, Zinyowera, and Moss, op. cit. note 27.
31. Bächler et al., op. cit. note 17; Gleick, op. cit. note 11; Flavin, op. cit. note 26; Myers, op. cit. note 2; Sam Howe Verhovek, "Wheat Farmers and Ranchers Are Ruined," *New York Times*, May 20, 1996.
32. Mexico from Thomas F. Homer-Dixon, "Environmental Scarcities and Violent Conflict: Evidence from Cases," *International Security*, Summer 1994; Myers, op. cit. note 2.

CHAPTER 3. Conflict Over the Environment

1. Volker Böge, "Bougainville: A 'Classical' Environmental Conflict?" Occasional Paper No. 3, Environment and Conflicts Project (ENCOP), Bern, Switzerland, October 1992.
2. Jörg Calließ, ed., *Treiben Umweltprobleme in Gewaltkonflikte?* Loccumer Protokolle 21/94 (Rehburg-Loccum, Germany: Evangelische Akademie Loccum, 1995); Günther Bächler, "Desertification and Conflict: The Marginalization of Poverty and of Environmental Conflicts," Occasional Paper No. 10, ENCOP, Bern, Switzerland, March 1994.
3. Günther Bächler et al., "Konfliktursache Umweltzerstörung," *Der Überblick*, March 1994.
4. All four factors are discussed in Thomas F. Homer-Dixon, "Environmental Scarcities and Violent Conflict: Evidence from Cases," *International Security*, Summer 1994.
5. Thomas F. Homer-Dixon, "On the Threshold: Environmental Changes as Causes of Acute Conflict," *International Security*, Fall 1991.
6. Volker Böge, "Anmerkungen und Überlegungen für die Weiterarbeit," in Calließ, op. cit. note 2; Homer-Dixon, op. cit. note 4.
7. Böge, op. cit. note 1.
8. Ibid.
9. Ibid.; mixture of fighting and cease-fires from Trevor Findlay, "Armed Conflict Prevention, Management and Resolution," in Stockholm International Peace Research Institute, *SIPRI Yearbook 1995: Armaments, Disarmament and International Security* (New York: Oxford University Press, 1995), from United Na-

tions, "Secretary-General Names Francesc Vendrell Envoy to All-Bougainvillean Talks to Be Held in Cairns, 15–19 December," press release, December 13, 1995, and from United Nations, "Secretary-General Concerned by Lifting of Cease-Fire on Bougainville," press release, March 22, 1996.

10. Peter B. Okoh, "Schutzpatronin der Ölkonzerne," *Der Überblick*, March 1994; Steve Kretzmann, "Nigeria's 'Drilling Fields': Shell Oil's Role in Repression," *Multinational Monitor*, January/February 1995.
11. Okoh, op. cit. note 10; Kretzmann, op. cit. note 10; "Nigeria: Erdölförderung zerstört Land," *Entwicklung + Zusammenarbeit*, August 1995.
12. Okoh, op. cit. note 10; Kretzmann, op. cit. note 10.
13. Kretzmann, op. cit. note 10.
14. Okoh, op. cit. note 10; Geraldine Brooks, "Shell's Nigerian Fields Produce Few Benefits for Region's Villagers," *Wall Street Journal*, May 6, 1994; Rob Nixon, "The Oil Weapon" (op-ed), *New York Times*, November 17, 1995; Howard F. French, "Nigeria Executes Critic of Regime; Nations Protest," *New York Times*, November 11, 1995; Rose Umoren, "Nigeria: Regime Hires Another Lobbyist," InterPress Service, December 6, 1995, on the APC electronic conference igc:africa.news on December 9, 1995.
15. Friends of the Earth, "Shell Disregards Tense Situation in Nigeria and Announces Plans to Go Ahead with Gas Project," press release, November 15, in the APC electronic conference igc:env.oil; Nixon, op. cit. note 14; "Shell Game in Nigeria" (editorial), *New York Times*, December 3, 1995.
16. Glenn Switkes, "Industry Out of Control" (letter to the editor), *New York Times*, November 17, 1995; Marcus Colchester, "Forest Conflicts Reach Flashpoint in Suriname," Third World Network Features, in the APC electronic conference igc:twn.features, October 10, 1995; Anthony de Palma, "In Suriname's Rain Forests, a Fight over Trees vs. Jobs," *New York Times*, September 4, 1995; *Development News 27.11*, in the APC electronic conference igc:hrnet.develop, November 29, 1995.
17. Number of shared rivers from "Environment and Conflict," Earthscan Briefing Document 40, International Institute for Environment and Development, London, November 1984; external flow from Peter H. Gleick, "Water and Conflict," in Occasional Paper No. 1, Project on Environmental Change and Acute Conflict, American Academy of Arts and Sciences and University of Toronto, September 1992; Table 3–1 is based on Michael Renner, *National Security: The Economic and Environ-*

mental Dimensions, Worldwatch Paper 89 (Washington, D.C.: Worldwatch Institute, May 1989), on Gleick, op. cit. in this note, on Günther Bächler, "The Anthropogenic Transformation of the Environment: A Source of War?" in Kurt R. Spillmann and Günther Bächler, eds., "Environmental Crisis: Regional Conflicts and Ways of Cooperation," Occasional Paper No. 14, ENCOP, Bern, Switzerland, September 1995, and on Andrew Nette, "Sharing the Mekong Creates Problems," Third World Network Features, Malaysia, on the APC electronic conference igc:twn.features on March 27, 1996.

18. Gleick, op. cit. note 17.
19. Israeli dependence from Miriam R. Lowi, "West Bank Water Resources and the Resolution of Conflict in the Middle East," in Occasional Paper No. 1, op. cit. note 17; reduction in Palestinian farmland from Günter Bächler et al., *Umweltzerstörung: Krieg oder Kooperation?* (Münster, Germany: Agenda Verlag, 1993); water in the context of Israeli-Arab peace is discussed in Stephan Libiszewski, "Water Disputes in the Jordan Basin Region and Their Role in the Resolution of the Arab-Israeli Conflict," Occasional Paper No. 13, ENCOP, Bern, Switzerland, August 1995, and in Spillmann and Bächler, op. cit. note 17.
20. Nile from Sandra Postel, "Forging a Sustainable Water Strategy," in Lester R. Brown et al., *State of the World 1996* (New York: W.W. Norton & Company, 1996); Euphrates from Udo Steinbach, "Wasser als Grund und Mittel internationaler Konflikte. Beispiel: Die Euphrat-Tigris Region," in Calließ, op. cit. note 2.
21. M. Abdul Hafiz and Nahid Islam, "Environmental Degradation and Intra/Interstate Conflicts in Bangladesh," Occasional Paper No. 10, ENCOP, Bern, Switzerland, May 1993.
22. Ibid.; Postel, op. cit. note 20; Sanjoy Hazarika, "Bangladesh and Assam: Land Pressures, Migration and Ethnic Conflict," in Occasional Paper No. 3, Project on Environmental Change and Acute Conflict, American Academy of Arts and Sciences and University of Toronto, March 1993; "Water in South Asia: Ganges River Fact Sheet," Southern Asian Institute, Columbia University, New York, May 1995.
23. Hafiz and Islam, op. cit. note 21; "Trouble in South Asia over Ganges Waters," *Development + Cooperation*, July/August 1995; saline water intrusion from Postel, op. cit. note 20.
24. Hafiz and Islam, op. cit. note 21; "Trouble in South Asia," op. cit. note 23; rice farmers and cost of crop loss from Ahmed Fazl, "Millions of Bangladeshis Threatened as the Ganges Dries up," IA News, in the APC electronic conference igc:alt.india.prog on May 2, 1995.

25. Hafiz and Islam, op. cit. note 21.
26. Sandra Postel, *Last Oasis: Facing Water Scarcity* (New York: W.W. Norton & Company, 1992); Marlise Simmons, "Brought to Knees, Arid Land Prays," *New York Times*, December 14, 1995; Stefan Klötzli, "The Water and Soil Crisis in Central Asia—A Source for Future Conflicts?" Occasional Paper No. 11, ENCOP, Bern, Switzerland, May 1994.
27. India from Catherina Hinz, "Die Kaveri-Kontroverse," *Der Überblick*, March 1994; Associated Press, "Mexico State Casts Water War," January 22, 1996, in CompuServe's AP Online.
28. Beijing diversion from Postel, op. cit. note 20; Vaclav Smil, "Environmental Change as a Source of Conflict and Economic Losses in China," in Occasional Paper No. 2, Project on Environmental Change and Acute Conflict, American Academy of Arts and Sciences and University of Toronto, December 1992.
29. Smil, op. cit. note 28.
30. Gleick, op. cit. note 17.
31. Displacements from "What Future for Large Dams?" *Development + Cooperation*, November/December 1995; World Bank study from Hal Kane, *The Hour of Departure: Forces That Create Refugees and Migrants*, Worldwatch Paper 125 (Washington, D.C.: Worldwatch Institute, June 1995).
32. Three Gorges from Aaron Sachs, *Eco-Justice: Linking Human Rights and the Environment*, Worldwatch Paper 127 (Washington, D.C.: Worldwatch Institute, December 1995), and from Pratap Chatterjee, "U.S. Companies Apply for Funding for Three Gorges Dam," InterPress Service, in the APC electronic conference igc:env.dams on December 6, 1995.
33. Volker Böge, "Das Sardar-Sarovar-Projekt an der Narmada in Indien—Gegenstand ökologischen Konflikts," Occasional Paper No. 8, ENCOP, Bern, Switzerland, June 1993.
34. Ibid.; "Water in South Asia: Narmada River Fact Sheet," Southern Asian Institute, Columbia University, New York, revised version, May 1995.
35. Böge, op. cit. note 33.
36. Ibid.
37. Ibid.; Narmada Bachao Andolan, "The Narmada Struggle: International Campaign After the World Bank Pullout," in the APC electronic conference igc:dev.worldbank on October 6, 1995.
38. Robert A. Hutchison, ed., *Fighting for Survival: Insecurity, People and the Environment in the Horn of Africa* (Gland, Switzerland: World Conservation Union, 1991); Norman Myers, *Ultimate Security: The Environmental Basis of Political Stability* (New York:

W.W. Norton & Company, 1993).

39. Bächler, op. cit. note 2; Astri Suhrke, "Pressure Points: Environmental Degradation, Migration and Conflict," in Occasional Paper No. 3, op. cit. note 22.
40. Suhrke, op.cit. note 39.
41. Mohamed Suliman, "Civil War in Sudan: The Impact of Ecological Degradation," Occasional Paper No. 4, ENCOP, Bern, Switzerland, December 1992.
42. Ibid.
43. Ibid.
44. Ibid.
45. Ibid.; Hutchison, op. cit. note 38; deaths and displacement from *Scottish Sudan News*, No. 23, December 1995, in the APC electronic conference igc:africa.horn on December 10, 1995.
46. Bächler, op. cit. note 2; Mohamed Suliman, "Conflicts of the Sudan," in Calließ, op. cit. note 2; Suhrke, op. cit. note 39; Hutchison, op. cit. note 38.
47. David E. Pitt, "U.N. Envoys Fear New Cod Wars as Fish Dwindle," *New York Times*, March 20, 1994.
48. Ibid.; West Africa from Deborah Cramer, "Troubled Waters," *Atlantic Monthly*, June 1995.
49. "Canada Condemned for Seizing Boat in Fishing Row," *Financial Times*, March 11/12, 1995; "EU Freezes Links with Canada Over Arrest of Trawler," *Financial Times*, March 14, 1995.
50. James Harding and Deborah Hargreaves, "Fish Knives Out in Defence of Canada's Turbot," *Financial Times*, March 11/12, 1995; Clyde H. Farnsworth, "Cod Are Almost Gone and a Culture Could Follow," *New York Times*, May 14, 1994; Clyde H. Farnsworth, "Canadians Cut the Nets of Spain Ship," *New York Times*, March 28, 1995; Clyde H. Farnsworth, "North Atlantic Fishing Pact Could Become World Model," *New York Times*, April 17, 1995.
51. Eileen Claussen, "Environment and Security: The Challenge of Integration," in *Environmental Change and Security Project Report* (Woodrow Wilson Center, Washington, D.C.), Spring 1995; "Practicing Piscacide: The Great, Global Fish Massacre," *Global Report* (Center for War/Peace Studies), Winter-Spring 1995; Bächler et al., op. cit. note 19; Pitt, op. cit. note 47.
52. Barbara Crossette, "Treaty Proposal to Curtail Overfishing is Approved at U.N.," *New York Times*, August 5, 1995; United Nations, Conference on Straddling and Highly Migratory Fish Stocks, "Conference on Straddling Fish Stocks Adopts Historic Agreement on Conservation of Depleted Stocks and Management of High-Seas Fisheries," press release, New York, August 4, 1995.

53. Matthew Gianni, "Greenpeace Analysis of the United Nations Treaty for the Conservation of Straddling Fish Stocks and Highly Migratory Fish Stocks," Greenpeace USA, September 1995, in the APC electronic conference igc:env.marine.

CHAPTER 4. Inequality and Insecurity

1. Somavia quoted in Barbara Crossette, "Despite the Risks, the U.N. Plans a World Conference on Poverty," *New York Times*, January 23, 1995.
2. Calvin Sims, "Workers Bitter at Pay and Privatization Tie Up Bolivian Capital," *New York Times*, March 28, 1996; lower-paid jobs filled by formerly middle-class people from United Nations, *World Social Situation in the 1990s* (New York: United Nations, 1994), and from Nathaniel C. Nash, "Latin Economic Speedup Leaves Poor in the Dust," *New York Times*, September 7, 1994.
3. Economic growth from Nash, op. cit. note 2; capital inflow from "New Debt Crisis in Latin America?," *Development + Cooperation*, May/June 1995.
4. Comparison of income inequality with pre-debt crisis situation from U.N. Economic Commission for Latin America and the Caribbean, *Social Panorama of Latin America 1994* (Santiago, Chile: 1994); U.N. poverty forecast and data on homeless from Nash, op. cit. note 2; regional poverty rate from "Mexico and Latin America: Poverty and Integration," *NAFTA and Inter-American Trade Monitor* (Institute for Agriculture and Trade Policy), December 1995 (based on a study by the Institute for Economic Research at the National Autonomous University of Mexico); Chile from "Oxfam Challenges the World Bank's Policies on Labour Market Deregulation," in the APC electronic conference igc:econ.saps on April 1, 1996; Nash, op. cit. note 2.
5. African share of foreign direct investment from Chakravarti Raghavan, "TNCs Control Two-Thirds of World Economy," Third World Network Features, in the APC electronic conference igc:twn.features on January 24, 1996; African loss under the Uruquay Round from Gary Gardner, "World Trade Climbing," in Lester R. Brown, Nicholas Lenssen, and Hal Kane, *Vital Signs 1995* (New York: W.W. Norton and Company, 1995).
6. North-South gap from U.N. Development Programme (UNDP), *Human Development Report 1995* (New York: Oxford University Press, 1995); Latin America most unequal from *NAFTA and Inter-American Trade Monitor*, op. cit. note 4; rich-

est 5 percent gaining at expense of poorest 75 percent from International Labour Organisation (ILO), *World Labour Report 1993* (Geneva: 1993); Table 4–1 based on UNDP, op. cit. in this note, and on David Dembo and Ward Morehouse, *The Underbelly of the U.S. Economy* (New York: Apex Press, 1995).

7. UNDP, *Human Development Report 1991* (New York: Oxford University Press, 1991); Table 4–2 based on various editions of the *Human Development Report*, on U.N. Department of Public Information (UNDPI) factsheets: "The Faces of Poverty" (March 1996), "The Geography of Poverty" (March 1996), and "Poverty: Casting Long Shadows" (February 1996), on U.N. Research Institute for Social Development (UNRISD), *States of Disarray: The Social Effects of Globalization* (Geneva: 1995), and on United Nations, "Backgrounder—Global Report on Human Settlements Reveals: 500 Million Homeless or Poorly Housed in Cities Worldwide," in the APC electronic conference igc:unic.news on February 9, 1996.
8. Nash, op. cit. note 2.
9. Rich-poor income gap from UNDP, *Human Development Report 1994* (New York: Oxford University Press, 1994), from UNDPI, "Poverty and Development: An (Im)balance Sheet" (February 1996), and from UNDPI, "The Geography of Poverty," op. cit. note 7; billionaires from Marcos Arrudal, "Education for Whose Development?" in the APC conference igc:econ.saps on November 18, 1995.
10. Number of poor people in western industrial countries from UNDP, op. cit. note 6; formerly Communist industrial countries from UNDP, op. cit. note 7; increase in inequality in rich nations from United Nations, op. cit. note 2; United Kingdom from UNRISD, op. cit. note 7; United States calculated from Dembo and Morehouse, op. cit. note 6.
11. UNDP, op. cit. note 6; Kerala from Arjun Makhijani, *From Global Capitalism to Economic Justice* (New York: Apex Press, 1992); Marc L. Miringhoff, *1995 Index of Social Health: Monitoring the Social Well-Being of the Nation* (Tarrytown, N.Y.: Fordham Institute for Innovation in Social Policy, 1995). This index combines data on, among others, infant mortality, teen drug abuse and suicide, unemployment, wages, health insurance coverage, poverty rates, rich-poor gaps, and access to affordable housing.
12. Mexico from Norman Myers, *Ultimate Security: The Environmental Basis of Political Stability* (New York: W.W. Norton & Company, 1993).
13. Table 4–3 based on ibid., on UNDPI, "The Geography of Poverty," op. cit. note 7, and on Myriam Vander Stichele, "Trade

Liberalization—The Other Side of the Coin," *Development + Cooperation*, January/February 1996; Sudan surplus from John Prendergast, "Greenwars in Sudan," *Center Focus*, July 1992; landless trends from Alan B. Durning, *Poverty and the Environment: Reversing the Downward Spiral*, Worldwatch Paper 92 (Washington, D.C.: Worldwatch Institute, November 1989).

14. For data on debt, commodity prices, and terms of trade, see International Monetary Fund, *International Financial Statistics Yearbook 1995* (Washington, D.C.: 1995); World Bank, *Commodity Trade and Price Trends, 1989–91* (Baltimore, Md.: John Hopkins University Press, 1993); World Bank, *World Debt Tables 1994–1995*, Vol. 1 (Washington, D.C.: World Bank, 1994).
15. Number of countries signed on to structural adjustment programs from Carlos Heredia and Steve Hellinger, "Getting to the Root of the Mexico Crisis," in the APC electronic conference igc:twn.features on September 25, 1995; UNRISD, op. cit. note 7.
16. Vicious cycle from U.N. Conference on Trade and Development, *The Least Developed Countries 1993–1994 Report* (New York: United Nations, 1994); fall in sub-Saharan Africa investment from ILO, op. cit. note 6; debt development from World Bank, *World Debt Tables 1994–1995*, op. cit. note 14, and from Hal Kane, "Third World Debt Rising Slowly," in Lester R. Brown, Hal Kane, and Ed Ayres, *Vital Signs 1993* (New York: W.W. Norton & Company, 1993); Walden Bello with Shea Cunningham and Bill Rau, *Dark Victory: The United States, Structural Adjustment and Global Poverty* (London: Pluto Press, 1994); African debt servicing from Bread for the World, "Easing Africa's Debt Burden," in the APC electronic conference igc:africa.news on March 6, 1996.
17. Samuel A. Morley, *Poverty and Inequality in Latin America: Past Evidence, Future Prospects*, Policy Essay No. 13 (Washington, D.C.: Overseas Development Council, 1994); sub-Saharan Africa from Bello, op. cit. note 16.
18. Anthony de Palma, "In Mexico, Hunger for Poor and Middle-Class Hardship," *New York Times*, January 15, 1995; drop in real wages from Carlos Heredia and Mary Purcell, "Structural Adjustment in Mexico: The Root of the Crisis," from Karen Hansen-Kuhn, "Structural Adjustment in Nicaragua: Sapping the Social Fabric," and from Karen Hansen-Kuhn, "Structural Adjustment in Costa Rica: Sapping the Economy," all in the APC electronic conference igc:econ.saps on March 7, 1995.
19. De Palma, op. cit. note 18; Heredia and Purcell, op. cit. note 18; Hansen-Kuhn, "Structural Adjustment in Nicaragua," op.

cit. note 18; Hansen-Kuhn, "Structural Adjustment in Costa Rica," op. cit. note 18.

20. Michel Chossudovsky, "The Causes of Global Famine," in the APC electronic conference igc:dev.worldbank on October 16, 1995; UNDP, op. cit. note 9.
21. India from Sandra Postel, "Forging a Sustainable Water Strategy," in Lester R. Brown et al., *State of the World 1996* (New York: W.W. Norton and Company, 1996); Colombia from Ekkehard Launer, *Zum Beispiel Blumen* (Göttingen, Germany: Lamuv Verlag, December 1994); Senegal and Mali from Chossudovsky, op. cit. note 20.
22. Threat to Brazilian peasants from Vander Stichele, op. cit. note 13; Brazilian imports from U.S. Department of Agriculture, Economic Research Service, "Production, Supply, and Distribution" (electronic database), Washington, D.C., November 1995.
23. Ray Marshall, "The Global Jobs Crisis," *Foreign Policy*, Fall 1995; Richard J. Barnet, "Lords of the Global Economy," *The Nation*, December 19, 1994; disparity in growth of GNP and employment from Hal Kane, *The Hour of Departure: Forces That Create Refugees and Migrants*, Worldwatch Paper 125 (Washington, D.C.: Worldwatch Institute, June 1995).
24. UNRISD, op. cit. note 7; United Nations, op. cit. note 2; labor pool from Barnet, op. cit. note 23; skilled labor in danger from Keith Bradsher, "Skilled Workers Watch Their Jobs Migrate Overseas," *New York Times*, August 28, 1995, and from Louis Uchitelle, "U.S. Corporations Expanding Abroad at a Quicker Pace," *New York Times*, July 25, 1994.
25. Declining real U.S. wages from Dembo and Morehouse, op. cit. note 6; U.S. job statistics from Louis Uchitelle and N.R. Kleinfield, "On the Battlefields of Business, Millions of Casualties," *New York Times*, March 3, 1996; poverty-level wages from Barnet, op. cit. note 23.
26. Rejection of cheap labor strategy from Marshall, op. cit. note 23; unemployment rates from ILO, *World Labour Report 1995* (Geneva: 1995), and from Associated Press, "G-7 Nations Fail on Jobs Plan," in CompuServe's AP Online on April 2, 1996; discussion of the measurement of U.S. unemployment from Dembo and Morehouse, op. cit. note 6.
27. See, for example, Uchitelle and Kleinfield, op. cit. note 25 (the lead article in a seven-piece *New York Times* series on "The Downsizing of America").
28. Sanford A. Marcus, "Downsizing: The Trashing of America's Soul" (letter to the editor), *New York Times*, March 8, 1996.
29. Global wage differentials from World Bank, *World Development*

Report 1995 (New York: Oxford University Press, 1995).
30. Ibid.; ILO, op. cit. note 26.
31. Marshall, op. cit. note 23.
32. United Nations, op. cit. note 2; ILO, op. cit. note 6.
33. Youth unemployment rates from UNDPI, "The Faces of Poverty," op. cit. note 7, from ILO, op. cit. note 6, and from ILO, op. cit. note 26.
34. Under-15 population shares from Population Reference Bureau, "1995 World Population Data Sheet" (wallchart), Washington, D.C., May 1995; possible reactions from UNDPI, "The Faces of Poverty," op. cit. note 7, and from Peter Gizewski and Thomas Homer-Dixon, "Urban Growth and Violence: Will the Future Resemble the Past?," The Project on Environment, Population, and Security, American Association for the Advancement of Science and University of Toronto, 1995.
35. Projected growth of global labor force from Kane, op. cit. note 23; developing countries' annual job creation needs from United Nations, op. cit. note 2.
36. UNRISD, op. cit. note 7.
37. Ibid.

Chapter 5. People on the Move

1. Sanjoy Hazarika, "Bangladesh and Assam: Land Pressures, Migration and Ethnic Conflict," in Occasional Paper No. 3, Project on Environmental Change and Acute Conflict, American Academy of Arts and Sciences and University of Toronto, March 1993; Sanjoy Hazarika, "Auf der Suche nach einem Leben in Würde," *Der Überblick*, March 1994.
2. Hazarika, "Bangladesh and Assam," op. cit. note 1; Astri Suhrke, "Pressure Points: Environmental Degradation, Migration and Conflict," in Occasional Paper No. 3, op. cit. note 1.
3. Hazarika, "Bangladesh and Assam," op. cit. note 1; Hazarika, "Auf der Suche," op. cit. note 1.
4. U. N. Research Institute for Social Development, *States of Disarray: The Social Effects of Globalization* (Geneva: 1995); Kenneth B. Noble, "Videotape of Beating by 2 Deputies Jolts Los Angeles," *New York Times*, April 3, 1996; Paul J. Smith, "Anti-Immigrant Xenophobia Around the World," *International Herald Tribune*, February 14, 1996, in the APC electronic conference igc:hrnet.racism on March 3, 1996.
5. Table 5–1 based on World Bank, *World Development Report 1995* (New York: Oxford University Press, 1995), and on U.N. Department for Economic and Social Information and Policy

Analysis, Population Division, "International Migration Policies 1995" (wallchart), United Nations, New York, December 1995.

6. U. N. High Commissioner for Refugees (UNHCR), *The State of the World's Refugees 1995* (New York: Oxford University Press, 1995); Hal Kane, *The Hour of Departure: Forces That Create Refugees and Migrants*, Worldwatch Paper 125 (Washington, D.C.: Worldwatch Institute, June 1995); Hal Kane, "Refugees on the Rise Again," in Lester R. Brown, Christopher Flavin, and Hal Kane, *Vital Signs 1996* (New York: W.W. Norton & Company, 1996).
7. Table 5–2 based on UNHCR, op. cit. note 6, and on Population Reference Bureau (PRB), "1995 World Population Data Sheet" (wallchart), Washington, D.C., May 1995.
8. UNHCR, op. cit. note 6.
9. PRB, op. cit. note 7; Ngara from Paul F. Macek, "A Tragic Outpouring: Rwandan Refugees, the Environment, and Security," Winston Foundation for World Peace, Washington, D.C., 1995.
10. Table 5–3 based on U.S. Committee for Refugees, *World Refugee Survey 1995* (Washington, D.C.: 1995), and on PRB, op. cit. note 7.
11. Liberia and Sierra Leone from UNHCR, op. cit. note 6.
12. Kane, *The Hour of Departure*, op. cit. note 6; UNHCR, op. cit. note 6; Table 5–4 based on U.N. Department for Economic and Social Information and Policy Analysis, op. cit. note 5; International Labour Organisation from UNHCR, op. cit. note 6.
13. Suhrke, op. cit. note 2.
14. Kane, *The Hour of Departure*, op. cit. note 6.
15. Ibid.
16. Jodi L. Jacobson, *Environmental Refugees: A Yardstick of Habitability*, Worldwatch Paper 86 (Washington, D.C.: Worldwatch Institute, November 1988); Kane, *The Hour of Departure*, op. cit. note 6.
17. Norman Myers, *Ultimate Security: The Environmental Basis of Political Stability* (New York: W.W. Norton & Company, 1993); Robert A. Hutchison, ed., *Fighting for Survival: Insecurity, People and the Environment in the Horn of Africa* (Gland, Switzerland: World Conservation Union, 1991).
18. Thomas F. Homer-Dixon, "On the Threshold: Environmental Changes as Causes of Acute Conflict," *International Security*, Fall 1991; Günter Bächler et al., *Umweltzerstörung: Krieg oder Kooperation?* (Münster, Germany: Agenda Verlag, 1993).
19. Vaclav Smil, "Environmental Change as a Source of Conflict and Economic Losses in China," in Occasional Paper No. 2,

Project on Environmental Change and Acute Conflict, American Academy of Arts and Sciences and University of Toronto, December 1992; Vaclav Smil, "China's Environmental Refugees: Causes, Dimensions and Risks of an Emerging Problem," in Kurt R. Spillmann and Günther Bächler, eds., "Environmental Crisis: Regional Conflicts and Ways of Cooperation," Occasional Paper No. 14, Environment and Conflicts Project (ENCOP), Bern, Switzerland, September 1995.

20. Smil, "Environmental Change," op. cit. note 19; Smil, "China's Environmental Refugees," op. cit. note 19.
21. Smil, "Environmental Change," op. cit. note 19; Jack A. Goldstone, "Imminent Political Conflicts Arising from China's Environmental Crises," in Occasional Paper No. 2, op. cit. note 19.
22. Jacobson, op. cit. note 16; share of population living near coast from Chapter 12, "Human Settlement in a Changing Climate: Impacts and Adaptation," in Robert T. Watson, Marufu C. Zinyowera, and Richard H. Moss, eds, *Climate Change 1995: Impacts, Adaptations and Mitigation of Climate Change: Scientific-Technical Analyses* (Cambridge: Cambridge University Press, for the Intergovernmental Panel on Climate Change, 1996).
23. Jacobson, op. cit. note 16; M. Abdul Hafiz and Nahid Islam, "Environmental Degradation and Intra/Interstate Conflicts in Bangladesh," Occasional Paper No. 6, ENCOP, Bern, Switzerland, May 1993.
24. Chapter 9, "Coastal Zones and Small Islands," in Watson, Zinyowera, and Moss, op. cit. note 22.
25. Chapter 13, "Agriculture in a Changing Climate: Impacts and Adaptation," in Watson, Zinyowera, and Moss, op. cit. note 22; Myers, op. cit. note 17.
26. Worldwatch extrapolation of UNHCR data reported in Macek, op. cit. note 9.
27. U.N. General Assembly, Fiftieth Plenary, "Seven Draft Texts Addressing Humanitarian Assistance Issue Introduced in General Assembly," press release, New York, November 27, 1995; UNHCR, op. cit. note 6.
28. UNHCR, "$70.5 Million Needed to Repair Damage by Rwandan Refugees," press release, Geneva, January 24, 1996; other impacts from Hutchison, op. cit. note 17.
29. UNHCR, op. cit. note 6.
30. Myron Weiner, *The Global Migration Crisis: Challenge to States and to Human Rights* (New York: Harper Collins, 1995), as cited in UNHCR, op. cit. note 6; Matthew Connelly and Paul Kennedy, "Must It Be the Rest Against the West?" *Atlantic Monthly*, December 1994.

31. UNHCR, op. cit. note 6.
32. Macek, op. cit. note 9.
33. UNHCR, op. cit. note 6; Steven Greenhouse, "Aristide to End Accord that Allows U.S. to Seize Refugee Boats," *New York Times*, April 8, 1994.
34. UNHCR, op. cit. note 6.

CHAPTER 6. Vicious Circles: Two Case Studies

1. Donatella Lorch, "Rwandan Refugees Describe Horrors After a Bloody Trek," *New York Times*, May 3, 1994; Raymond Bonner, "Rwandans in Death Squad Say Choice Was Kill or Die," *New York Times*, August 14, 1994; "Rwanda Killers Leave a Village of the Dead," *New York Times*, May 14, 1994.
2. Donatella Lorch, "In the Upheaval in Rwanda, Few Answers Yet," *New York Times*, May 5, 1994; Milton Leitenberg, "Anatomy of a Massacre" (op-ed), *New York Times*, July 31, 1994.
3. James Murray, "Rwanda's Bloody Roots" (op-ed), *New York Times*, September 3, 1994; Valerie Percival and Thomas Homer-Dixon, "Environmental Scarcity and Violent Conflict: The Case of Rwanda," The Project on Environment, Population, and Security, American Association for the Advancement of Science and University of Toronto, June 1995.
4. Percival and Homer-Dixon, op. cit. note 3.
5. Günther Bächler, "Welche Rolle spielt Ökologie als Ursache und Medium von (zukünftigen) Gewaltkonflikten im internationalen System?," in Jörg Calließ, ed., *Treiben Umweltprobleme in Gewaltkonflikte?*, Loccumer Protokolle 21/94 (Rehburg-Loccum, Germany: Evangelische Akademie Loccum, 1995); Human Rights Watch, *Slaughter Among Neighbors: The Political Origins of Communal Violence* (New Haven, Conn.: Yale University Press, 1995).
6. Peter Molt, "Ein Produkt der Kolonialherrschaft," *Frankfurter Rundschau*, June 20, 1994; Percival and Homer-Dixon, op. cit. note 3; Human Rights Watch, op. cit. note 5.
7. Percival and Homer-Dixon, op. cit. note 3.
8. Ibid.
9. Molt, op. cit. note 6; Population Reference Bureau, "1995 World Population Data Sheet" (wallchart), Washington, D.C., May 1995.
10. World Bank, *World Development Report 1995* (New York: Oxford University Press, 1995).
11. Area under cultivation from U.S. Department of Agriculture (USDA), Economic Research Service, "Production, Supply, and Distribution" (electronic database), Washington, D.C., Novem-

ber 1995; marginal land from Bächler, op. cit. note 5; conversion of pastureland from Lloyd Timberlake, *Africa in Crisis* (London: Earthscan Publications, 1985).

12. Percival and Homer-Dixon, op. cit. note 3; grain area and yields from USDA, op. cit. note 11.
13. USDA, op. cit. note 11; declining self-sufficiency from Nikos Alexandratos, ed., *World Agriculture: Towards 2010. An FAO Study* (New York: John Wiley & Sons, 1995); Figure 6–1 from USDA, op. cit. note 11.
14. Percival and Homer-Dixon, op. cit. note 3; Frank Smyth, "Arms for Rwanda: Blood Money and Geopolitics," *The Nation*, May 3, 1994.
15. World Bank, *Commodity Trade and Price Trends, 1989–91* (Baltimore, Md.: Johns Hopkins University Press, 1993); World Bank, op. cit. note 10; World Bank, *World Debt Tables 1994–95*, Vol. 1 (Washington, D.C.: 1994); falling per capita gross domestic product from André Guichaoua, "Kein Staat wird allein zum Frieden finden," *Der Überblick,* March 1996.
16. Molt, op. cit. note 6; Percival and Homer-Dixon, op. cit. note 3; Lorch, op. cit. note 2.
17. Molt, op. cit. note 6; Percival and Homer-Dixon, op. cit. note 3.
18. Molt, op. cit. note 6; Percival and Homer-Dixon, op. cit. note 3; Human Rights Watch, op. cit. note 5.
19. Lorch, op. cit. note 2; Bonner, op. cit. note 1; Human Rights Watch, op. cit. note 5; Howard Adelman and Astri Suhrke, "Feilschen während Ruanda brennt," *Der Überblick,* March 1996.
20. Tim Golden, "Mexican Troops Battling Rebels; Toll at Least 56," *New York Times*, January 3, 1994; Tim Golden, "Mexican Copters Pursue Rebels; Death Toll in Uprising Is Put at 95," *New York Times*, January 6, 1994.
21. National resonance from Tom Barry, *Zapata's Revenge: Free Trade and the Farm Crisis in Mexico* (Boston, Mass.: South End Press, 1995); takeovers and scrutiny from Tim Golden, "'Awakened' Peasant Farmers Overrunning Mexican Towns," *New York Times*, February 9, 1994, and from George A. Collier with Elizabeth Lowery Quaratiello, *Basta! Land and the Zapatista Rebellion in Chiapas* (Oakland, Calif.: Food First Books, Institute for Food and Development Policy, 1994); initiation of peace talks from Anthony DePalma, "Mexico Leader Names Negotiator to Seek Political End to Uprising," *New York Times*, January 11, 1994, and from Anthony DePalma, "Mexico Orders Cease-Fire and Offers Rebels Amnesty," *New York Times*, January 13, 1994.

22. Thomas Benjamin, *A Rich Land, a Poor People: Politics and Society in Modern Chiapas* (Albuquerque: University of New Mexico Press, 1989); Chiapas's share of national outputs from Collier, op. cit. note 21, and from Philip Howard and Thomas Homer-Dixon, "Environmental Scarcity and Violent Conflict: The Case of Chiapas, Mexico," The Project on Environment, Population, and Security, American Association for the Advancement of Science and University of Toronto, January 1996; beef from Barry, op. cit. note 21; running water and literacy from "The Mexican Rebels' Impoverished Home," *New York Times*, January 9, 1994.
23. Collier, op. cit. note 21; Indian fatalities from Gustavo Esteva, "Mexican Indians Say No to Development," *The People-Centered Development Forum*, May 20, 1994.
24. Howard and Homer-Dixon, op. cit. note 22.
25. Ibid.; pastureland from Barry, op. cit. note 21.
26. Barry, op. cit. note 21; grain trends from USDA, op. cit. note 11, and from Alexandratos, op. cit. note 13.
27. Collier, op. cit. note 21; Lacandón population growth from Homero Aridjis, "Slaves and Guerrillas, Forests and Blood" (op-ed), *New York Times*, January 5, 1994.
28. Collier, op. cit. note 21; decline of Lacandón tree cover from Barry, op. cit. note 21; doubling of pastureland from Howard and Homer-Dixon, op. cit. note 22; border visibility from space from "The Mexican Rebels' Impoverished Home," op. cit. note 22.
29. Collier, op. cit. note 21; redistribution of land and reduction in landlessness from Barry, op. cit. note 21.
30. Collier, op. cit. note 21; Barry, op. cit. note 21.
31. Barry, op. cit. note 21; Andrew Reding, "Chiapas Is Mexico: The Imperative of Political Reform," *World Policy Journal*, Spring 1994.
32. Collier, op. cit. note 21; Barry, op. cit. note 21.
33. Collier, op. cit. note 21; removal of agricultural tariffs from Reding, op. cit. note 31.
34. Collier, op. cit. note 21.
35. Lack of credit access from Howard and Homer-Dixon, op. cit. note 22; INMECAFE from Collier, op. cit. note 21.
36. PRONASOL funds from U.N. Economic Commission for Latin America and the Caribbean, *Social Panorama of Latin America 1994* (Santiago, Chile: 1994); failure to win support in Chiapas from Collier, op. cit. note 21; World Bank from Barry, op. cit. note 21.
37. Collier, op. cit. note 21.
38. Reding, op. cit. note 31; Andrew Reding, "Human Rights,

Chiapas, Spring 1993" (op-ed), *New York Times*, January 7, 1994.
39. Collier, op. cit. note 21.
40. Reding, op. cit. note 31.
41. Ibid.; Jorge G. Castañeda, "Ferocious Differences," *Atlantic Monthly*, July 1995; Barry, op. cit. note 21.
42. Anthony DePalma, "Mexican Troops Seize Rebel Base But Don't Find Leader," *New York Times*, February 10, 1995; Julia Preston, "Mexico and Insurgent Group Reach Pact on Indian Rights," *New York Times*, February 15, 1996; Associated Press, "Govt, Zapatistas Reach Accord," in CompuServe's AP Online, February 16, 1996; Trina Kleist, "Zapatistas OK Postponing Talks," Associated Press, June 3, 1996, on CompuServe. A leaked internal memo of Chase Manhattan Bank stated: "While Chiapas, in our opinion, does not pose a fundamental threat to Mexican political stability, it is perceived to be so by many in the investment community. The government will need to eliminate the Zapatistas to demonstrate their effective control of the national territory and of security policy." The memo was reproduced in the APC electronic conference igc:reg.mexico on February 15, 1995.

CHAPTER 7. A Human Security Policy

1. United Nations, "Report of the World Summit for Social Development (Copenhagen, 6–12 March 1995)," New York, April 19, 1995.
2. Gareth Porter and Janet Welsh Brown, *Global Environmental Politics* (Boulder, Colo.: Westview Press, 1991).
3. Hilary F. French, "Forging a New Global Partnership," in Lester R. Brown et al., *State of the World 1995* (New York: W.W. Norton & Company, 1995); Anjali Acharya, "CFC Production Drop Continues," in Lester R. Brown, Christopher Flavin, and Hal Kane, *Vital Signs 1996* (New York: W.W. Norton & Company, 1996); Hilary F. French, "Environmental Treaties Grow in Number," in Lester R. Brown, Nicholas Lenssen, and Hal Kane, *Vital Signs 1995* (New York: W.W. Norton & Company, 1995).
4. French, "Forging a New Global Partnership," op. cit. note 3; Jens Mertens and Peter Mucke, "Special Session of the UN General Assembly 1997: Core Issues," as posted by NGO Forum on Environment and Development in the APC electronic conference igc:un.csd.general on May 11, 1996; Commission on Sustainable Development from International Institute for Sustainable Development, *Earth Negotiations Bulletin*, Vol. 5, No. 57, in the APC electronic conference igc:un.csd.general on May

5, 1996.

5. Michael Renner, *National Security: The Economic and Environmental Dimensions*, Worldwatch Paper 89 (Washington, D.C.: Worldwatch Institute, May 1989).
6. Aaron Sachs, "Upholding Human Rights and Environmental Justice," in Lester R. Brown et al., *State of the World 1996* (New York: W.W. Norton & Company, 1996); Deborah Cramer, "Troubled Waters," *Atlantic Monthly*, June 1995.
7. Sachs, op. cit. note 6.
8. Peter H. Sand, "International Cooperation: The Environmental Experience," in Jessica Tuchman Mathews, ed., *Preserving the Global Environment: The Challenge of Shared Leadership* (New York: W.W. Norton & Company, 1991).
9. Thirty-Percent Club from Renner, op. cit. note 5.
10. United Nations, op. cit. note 1.
11. Ibid.; International Institute for Sustainable Development, *Earth Negotiations Bulletin*, March 15, 1995.
12. U.N. Research Institute for Social Development, *States of Disarray: The Social Effects of Globalization* (Geneva: 1995); James Fallows, "How the World Works," *Atlantic Monthly*, December 1993.
13. Equipo Pueblo, *Mexico Update No. 70*, May 2, 1996, in the APC electronic conference igc:reg.mexico on May 2, 1996.
14. Michael Holman and Patti Waldmeir, "Poor Nations' Debt Burden Lies Heavily on World Bank Minds," *Financial Times*, March 14, 1996; "World Bank May Create Debt Fund," *Multilateral News*, September 27, 1995; Reuters, "British Charities Urge Debt Write-Off for Africa," on CompuServe on February 4, 1996.
15. Rose Umoren, "Firm Debt Plan Slips Target Date," InterPress Service, April 18, 1996, and Rose Umoren, "World Bank/IMF Meet Leaves Poor Countries Hanging," InterPress Service, April 23, 1996, both posted in the APC electronic conference igc:africa.news on April 30, 1996.
16. Alan B. Durning, *Poverty and the Environment: Reversing the Downward Spiral*, Worldwatch Paper 92 (Washington, D.C.: Worldwatch Institute, November 1989).
17. Diana Jean Schemo, "Brazilian Squatters Fall in Deadly Police Raid," *New York Times*, September 19, 1995; Diana Jean Schemo, "Brazil's Chief Acts to Take Land to Give to the Poor," *New York Times*, November 13, 1995; Diana Jean Schemo, "Violence Growing in Battle Over Brazilian Land," *New York Times*, April 21, 1996; Movimiento data from "Brazil: Agrarian Reform Proposed to End Violence," in *NAFTA & Inter-American Trade*

Monitor (Institute for Agriculture and Trade Policy), November 3, 1995, in the APC electronic conference igc:trade.news on November 3, 1995.

18. Hal Kane, "Microenterprise," *World Watch*, March/April 1996.
19. Ibid.; Durning, op. cit. note 16; Yunus quote from Patrick E. Tyler, "Star at Conference on Women: Banker Who Lends to the Poor," *New York Times*, September 14, 1995.
20. Tyler, op. cit. note 19.
21. Edward A. Gargan, "'People's Banks Help Rescue Poor Indonesians," *New York Times*, February 18, 1996; Durning, op. cit. note 16; number of people served by micro-credit institutions from Tyler, op. cit. note 19; examples of other initiatives, micro-credit summit, and World Bank from Kane, op. cit. note 18.
22. Stephan Schmidheiny, with the Business Council for Sustainable Development, *Changing Course* (Cambridge, Mass.: The MIT Press, 1992).
23. The issue of corporate charters in the U.S. context is discussed in Richard L. Grossman and Frank T. Adams, *Taking Care of Business: Citizenship and the Charter of Incorporation* (Cambridge, Mass.: Charter, Ink., 1993).
24. For endorsement of this point, see the "We Believe Statement" in President's Council on Sustainable Development, *Sustainable America* (Washington, D.C.: U.S. Government Printing Office, 1996).
25. Durning, op. cit. note 16; Aaron Sachs, *Eco-Justice: Linking Human Rights and the Environment*, Worldwatch Paper 127 (Washington, D.C.: Worldwatch Institute, December 1995).
26. Sachs, op. cit. note 25.
27. Commission on Global Governance, *Our Global Neighborhood* (New York: Oxford University Press, 1995); United Nations, *Directory of Non-Governmental Organizations and NGO's Representatives Associated with the Department of Public Information* (New York: U.N. Department of Public Information, 1994).
28. Earth Summit Watch, *Four in '94—Assessing National Actions to Implement Agenda 21: A Country-by-Country Report* (Washington, D.C.: 1994); Social Watch, *Social Watch: The Starting Point*, Trial Edition 1996 (Montevideo, Uruguay: 1996). Social Watch has a World Wide Web home page at http://www.chasque.apc.org/socwatch/; country reports are being posted on the APC electronic conference igc:socsummit.
29. Commission on Global Governance, op. cit. note 27; Michael Renner, "The Decline of Nations and the Future of the U.N." (editorial), *World Watch*, March/April 1996.

Chapter 8. Enhancing International Peace Capacity

1. Ruth Leger Sivard, *World Military and Social Expenditures 1993* (Washington, D.C.: World Priorities, 1993).
2. Michael Renner, "Environmental Dimensions of Disarmament and Conversion," in Kevin J. Cassidy and Gregory A. Bischak, eds., *Real Security: Converting the Defense Economy and Building Peace* (Albany: State University of New York Press, 1993).
3. The environment-related missions of the U.S. military are approvingly discussed by Kent Hughes Butts, "Why the Military is Good for the Environment," in Jyrki Käkönen, ed., *Green Security or Militarized Environment* (Aldershot, U.K.: Dartmouth Publishing Co., 1994).
4. Christopher Louise, "The Social Impacts of Light Weapons Availability and Proliferation," Discussion Paper No. 59, U.N Research Institute for Social Development, Geneva, March 1995.
5. See the chapters on arms transfers published annually in Stockholm International Peace Research Institute (SIPRI), *SIPRI Yearbook* (New York: Oxford University Press) for detailed discussion; Kalashnikovs from *International Security Digest*, November 1994.
6. Edward J. Laurance and Herbert Wulf, eds., *Coping with Surplus Weapons: A Priority for Conversion Research and Policy*, Brief 3 (Bonn, Germany: Bonn International Center for Conversion (BICC), 1995); Ksenia Gonchar and Peter Lock, "Small Arms and Light Weapons: Russia and the Former Soviet Union," in Jeffrey Boutwell, Michael T. Klare, and Laura W. Reed, eds., *Lethal Commerce: The Global Trade in Small Arms and Light Weapons* (Cambridge, Mass.: American Academy of Arts and Sciences, 1995); share of secondhand arms sales from Susanne Kopte, Michael Renner, and Peter Wilke, "The Cost of Disarmament: Dismantlement of Weapons and the Disposal of Military Surplus," *The Nonproliferation Review*, Winter 1996.
7. Laurance and Wulf, op. cit. note 6.
8. Licensed production trend from U.S. Congress, Office of Technology Assessment, *Global Arms Trade* (Washington, D.C.: U.S. Government Printing Office, 1991) and from SIPRI, op. cit. note 5; Jane's survey reported in Louise, op. cit. note 4.
9. Arms exports data derived from U.S. Arms Control and Disarmament Agency, *World Military Expenditures and Arms Transfers* (Washington, D.C.: U.S. Government Printing Office, various years); small arms transfers estimate from Louise, op. cit., note 4; Pentagon projection reported in *Arms Trade News*, March

1995; Defense Intelligence Agency from *Arms Sales Monitor*, July 20, 1995; cumulative past arms trade figure from Michael Renner, "Arms Trade Continues Decline," in Lester R. Brown, Hal Kane, and David Malin Roodman, *Vital Signs 1994* (New York: W.W. Norton & Company, 1994).

10. At a hearing of the House Foreign Affairs Committee on November 10, 1993, for instance, Under-Secretary of State Lynn Davis argued that "Arms sales are appropriate to responsible allies, and that is where our sales have been going"; *Arms Sales Monitor*, July 30, 1994. This argument is also enshrined in the administration's official policy statement on conventional arms transfers, an unclassified summary of which was released on February 17, 1995; see *Arms Sales Monitor*, March 20, 1995. U.S. arms transfers to conflict areas from William D. Hartung, *U.S. Weapons at War: Arms Deliveries to Regions of Conflict*, World Policy Papers (New York: World Policy Institute, 1995).
11. Chris Smith, "Light Weapons and Ethnic Conflict in South Asia," and Jo L. Husbands, "Controlling Transfers of Light Arms: Linkages to Conflict Processes and Conflict Resolution Strategies," both in Boutwell, Klare, and Reed, op. cit. note 6; Hartung, op. cit. note 10.
12. Vietnam from Husbands, op. cit. note 11; Central America from Daniel Garcia-Peña Jaramillo, "Light Weapons and Internal Conflict in Colombia," in Boutwell, Klare, and Reed, op. cit. note 6; Lebanon and Mozambique from Louise, op. cit. note 4, and from World Bank, "Demobilization and Reintegration of Military Personnel in Africa: The Evidence from Seven Country Case Studies," Discussion Paper, Africa Regional Series, Washington, D.C., October 1993.
13. A number of private research and advocacy groups have begun to track and study the trade in small arms. For example, the British American Security Council (BASIC) initiated a project that seeks to establish an international network of researchers; see BASIC, "Project on Light Weapons," informational flyer, April 1996. A first research result is a book: Jasjit Singh, ed., *Light Weapons and International Security* (Delhi: Indian Pugwash Society/BASIC, 1995).
14. For an overview, see Edward J. Laurance and Herbert Wulf with Joseph DiChiaro, III, *Conversion and the Integration of Economic and Security Dimensions*, BICC Report 1 (Bonn: BICC, 1995); lack of funding from Michael Renner, *Budgeting for Disarmament: The Costs of War and Peace*, Worldwatch Paper 122 (Washington, D.C.: Worldwatch Institute, November 1995).
15. El Salvador from Edward J. Laurance, "Surplus Weapons and

the Micro-Disarmament Process," Monterey Institute of International Studies, unpublished, November 1995; Mozambique from Kees Kingma, "Demobilisierung und Wiedereingliederung von Soldaten: Wichtige Schritte im Friedensprozess," in Hanne-Margret Birckenbach, Uli Jäger, and Christian Wellmann, eds., *Jahrbuch Frieden 1996* (München, Germany: Verlag C.H. Beck, 1995).

16. U.N. advisory missions from Swadesh Rana, "Small Arms and Intra-State Conflicts," Research Paper No. 34, U.N. Institute for Disarmament Research, Geneva, March 1995; General Assembly from Stephen W. Young, "The Fiftieth Session of the UN First Committee," *BASIC Reports*, December 1, 1995; survey from United Nations, "United Nations Global Study Begun on Civilian-Owned Firearms, Small Arms Trafficking, Firearms Regulations," press release, Vienna, December 22, 1995; South Asia from Smith, op. cit. note 11.
17. Francisco José Aguilar, ed., *Demobilization, Demilitarization, and Democratization in Central America* (San José, Costa Rica: Arias Foundation for Peace and Human Progress, 1994); Kees Kingma and Vanessa Sayers, *Demobilization in the Horn of Africa,* Brief 4 (Bonn, Germany: BICC, 1995).
18. Malcolm Chalmers et al., eds., *Developing the UN Register of Conventional Arms,* Bradford Arms Register Studies No. 4 (Bradford, U.K.: University of Bradford, 1994).
19. Ibid.; Edward J. Laurance and Tracy M. Keith, "The United Nations Register of Conventional Arms: On Course in its Third Year of Reporting," *The Nonproliferation Review,* Winter 1996; Susannah L. Dyer and Natalie J. Goldring, "US and Germany Dominate World Weapons Exports," *BASIC Reports*, November 20, 1995.
20. While an expansion of weapons categories included in the Register is desirable, there also is a need for consistency and continuity in the weapons categories and definitions; otherwise, comparisons over time will be impossible; Chalmers et al., op. cit. note 18. Regional registers proposed in "U.S. and Germany Dominate World Weapons Exports According to U.N. Arms Register," BASIC USA, press release, in the APC electronic conference igc:disarm.armstra on November 1, 1995.
21. Stephen Kinzer, "Nobel Peace Laureates Draft a Plan to Govern Arms Trade," *New York Times*, September 6, 1995; "Speech by Dr. Oscar Arias at Capitol Hill Symposium," Washington, D.C., December 15, 1995, in the APC electronic conference igc:disarm.armstra on December 19, 1995.
22. U.S. efforts from "Senate Funding Chief and House Foreign

Policy Activist Announce Bipartisan Campaign Against U.S. Arms Sales to Dictators," from "Republicans Save Clinton Arms Sales Policy from Solid Democratic Attack; 18–17 Defeat of 'No Arms to Dictators' Amendment to be Revisited on House Floor," and from "Democrats' Bid to Ban Arms Sales to Dictators Turned Back by House; Republicans Save Clinton Policy," all press releases from the Project on Demilitarization and Democracy in the APC electronic conference igc:disarm.armstra on February 2, 1995, May 12, 1995, and May 26, 1995; BASIC, "A European Code of Conduct on the Arms Trade," in the APC electronic conference igc:disarm.armstra on June 13, 1995.

23. The Year 2000 Campaign to Redirect World Military Spending to Human Development, "Campaign Statement as presented by Dr. Oscar Arias to U.N. Secretary-General Boutros Boutros-Ghali and at a Capitol Hill Symposium sponsored by Senator Mark O. Hatfield," December 1995.
24. Michael Renner, "Remaking U.N. Peacekeeping: U.S. Policy and Real Reform," Briefing Paper 17, National Commission for Economic Conversion and Disarmament, Washington, D.C., November 1995.
25. Erskine Childers, "UN Mechanisms and Capacities for Intervention," in Elizabeth G. Ferris, ed., *The Challenge to Intervene: A New Role for the United Nations?* (Uppsala, Sweden: Life and Peace Institute, 1992).
26. Michael Renner, *Critical Juncture: The Future of Peacekeeping*, Worldwatch Paper 114 (Washington, D.C.: Worldwatch Institute, May 1993).
27. Ibid.; Commission on Global Governance, *Our Global Neighborhood* (New York: Oxford University Press, 1995).
28. Peter Schlotter, "Von der KSZE zur OSZE: Marginalisierung oder neue Aufgaben?" in Birckenbach, Jäger, and Wellmann, op. cit. note 15.
29. Ibid.
30. Rolf Hofmeier, "State Disintegration in Africa," *Development + Cooperation*, May/June 1995.
31. Jane E. Holl, Executive Director, Carnegie Commission on Preventing Deadly Conflict, "Progress Report," Carnegie Corporation of New York, July 1995; Institute for Resource and Security Studies, "An Introduction to IRSS's Project on Preventive Diplomacy and National Security," Cambridge, Mass., November 1994; for activities of International Alert, London, see their newsletter.
32. Larry Diamond, *Promoting Democracy in the 1990s: Actors and Instruments, Issues and Imperatives* (New York: Carnegie Corporation of New York, 1995).
33. For more discussion of such a peace operations structure, see Renner, op. cit. note 24, and Renner, op. cit. note 26.

34. Early support for this idea was voiced by Edward C. Luck and Tobi Trister Gati, "Whose Collective Security?" *The Washington Quarterly*, Spring 1992.
35. Jonathan Dean, "A Stronger U.N. Strengthens America," *Bulletin of the Atomic Scientists*, March/April 1995.

CHAPTER 9. A Human Security Budget

1. Jessica Mathews, "Robbing Development to Pay for Disaster Relief" (op-ed), *Washington Post*, July 5, 1994.
2. U.N. High Commissioner for Refugees from Dutch Ministry of Foreign Affairs, *Humanitarian Aid Between Conflict and Development* (The Hague: 1993), and from Ruben P. Mendez, "Paying for Peace and Development," *Foreign Policy*, Fall 1995; World Food Programme from United Nations, "Pledging Conference for Development Activities Begins Two-day Session," press release, New York, November 1, 1994, and from United Nations, "Press Conference by Secretary-General Boutros-Ghali," Copenhagen, Denmark, press release, March 7, 1995.
3. UNICEF from "Trotting to the Rescue," *Economist*, June 25, 1994, and from UNICEF, *The State of the World's Children 1996* (New York: Oxford University Press, 1995); Organisation for Economic Co-operation and Development from Bernard Wood, Director of the Development Assistance Committee, remarks at the "Presentation of PRODERE External Evaluation Report," United Nations Headquarters, New York, February 6, 1995, and from Bernd Ludermann, "Entwicklungshilfe in der Not," *Der Überblick*, September 1995.
4. Michael Renner, *Budgeting for Disarmament: The Costs of War and Peace*, Worldwatch Paper 122 (Washington, D.C.: Worldwatch Institute, November 1994); Commission on Global Governance, *Our Global Neighborhood* (New York: Oxford University Press, 1995).
5. Table 9–1 based on Worldwatch database developed for Renner, op. cit. note 4, and on Greg Bischak, National Commission for Economic Conversion and Disarmament, Washington, D.C., various private communications, April 1996; need for twentyfold minimum increase in demining funds from United Nations General Assembly, "Assistance in Mine Clearance. Report of the Secretary-General," New York, September 4, 1994.
6. Training costs from Barry M. Blechman and J. Matthew Vaccaro, *Training for Peacekeeping: The United Nations' Role*,

Occasional Papers Series, Report No. 12 (Washington, D.C.: Henry L. Stimson Center, 1994); all other figures are Worldwatch estimates.

7. Carl Conetta and Charles Knight, *Vital Force: A Proposal for the Overhaul of the UN Peace Operations System and for the Creation of a UN Legion*, Project on Defense Alternatives Research Monograph 4 (Cambridge, Mass.: Commonwealth Institute, October 1995).
8. R&D estimate from Bonn International Center for Conversion (BICC), *Conversion Survey 1996: Global Disarmament, Demilitarization and Demobilization* (New York: Oxford University Press, 1996).
9. Table 9–2 based on Economists Allied for Arms Reduction, "What the World Wants—and How to Pay for It," undated fact sheet, on U.N. Development Programme (UNDP), *Human Development Report 1994* (New York: Oxford University Press, 1994), on United Nations, "UN Financial Crisis Affecting Operational Activities for Development, Says Secretary-General at Development Pledging Conference," press release, New York, November 1, 1995, on Patti L. Petesch, *North-South Environmental Strategies, Costs, and Bargains*, Policy Essay No. 5 (Washington, D.C.: Overseas Development Council, 1992), on Norman Myers, *Ultimate Security: The Environmental Basis of Political Stability* (New York: W.W. Norton & Company, 1993), on Andrew Jordan, "Financing the UNCED Agenda: The Controversy over Additionality," *Environment*, April 1994, and on Dragoljub Najman and Hans d'Orville, *Towards a New Multilateralism: Funding Global Priorities* (New York: Independent Commission on Population and Quality of Life, May 1995).
10. Fossil fuel costs calculated from British Petroleum, *BP Statistical Review of World Energy* (London: Group Media & Publications, 1995), and from International Monetary Fund, *International Financial Statistics* (Washington, D.C.: April 1996).
11. Military spending and R&D from BICC, op. cit. note 8; arms production from Elisabeth Sköns and Ksenia Gonchar, "Arms Production," in Stockholm International Peace Research Institute, *SIPRI Yearbook 1995* (New York: Oxford University Press, 1995).
12. Oscar Arias Sánchez, "A Global Demilitarization Fund," Special Contribution to UNDP, op. cit. note 9; Table 9–3 from Renner, op. cit. note 4.
13. Jordan, op. cit. note 9.
14. Hilary F. French, *Partnership for the Planet: An Environmental*

Agenda for the United Nations, Worldwatch Paper 126 (Washington, D.C.: Worldwatch Institute, July 1995); "Russia to Ask for Delay in Rules to Protect Ozone," *New York Times*, November 15, 1995.

15. French, op. cit. note 14; Jordan, op. cit. note 9.
16. UNDP, op. cit. note 9; Harriet Lamb, "British Government Found Guilty in Aid-for-Arms Scandal," *International Security Digest*, February 1995.
17. United Nations, "Pledging Conference for Development Activities," op. cit. note 2. The United States now contributes only 17 percent of worldwide development aid; Environmental and Energy Study Institute, "U.S. Development Assistance: A Visual Briefing," prepared for the International Development Conference, Washington, D.C., January 16–18, 1995; Hilary F. French, "Private Finance Flows to Third World," in Lester R. Brown, Christopher Flavin, and Hal Kane, *Vital Signs 1996* (New York: W.W. Norton & Company, 1996).
18. United Nations, "Report of the World Summit for Social Development (Copenhagen, 6–12 March 1995)," New York, April 19, 1995; International Institute for Sustainable Development, *Earth Negotiations Bulletin*, March 15, 1995.
19. UNDP, op. cit. note 9.
20. Data are calculated on the basis of the summary tables published in World Bank, *World Debt Tables 1994–95*, Vol. 1 (Washington, D.C.: 1994); debt reduction through swaps from Najman and d'Orville, op. cit. note 9.
21. World Bank, op. cit. note 20; sub-Saharan Africa from Julian Samboba, "Africa-Development: Debt Burden Growing at 'Alarming Rate'," InterPress Service, in the APC electronic conference igc:un.socsummit, May 10, 1995; Tanzania from U.N. Conference on Trade and Development, "UNCTAD IX Discusses Effects of Free Market Forces, Foreign Debt, Workers' Rights, in General Debate," press release, Midrand, South Africa, May 3, 1996.
22. Debt data from World Bank, op. cit. note 20.
23. Najman and d'Orville, op. cit. note 9.
24. James Tobin, "A Tax on International Currency Transactions," Special Contribution to UNDP, op. cit. note 9; variants of the proposal discussed by Najman and d'Orville, op. cit. note 9, and by Ruben P. Mendez, "Paying for Peace and Development," *Foreign Policy*, Fall 1995; Commission on Global Governance, op. cit. note 4.
25. Private foreign direct investment from French, op. cit. note 17.

Chapter 10. A Global Partnership for Human Security?

1. Robert D. Kaplan, "The Coming Anarchy," *Atlantic Monthly*, February 1994.
2. Robert Kaplan, "'The Coming Anarchy' and the Nation-State Under Siege," *Peaceworks* (U.S. Institute of Peace, Washington, D.C.), August 1995.
3. Shevardnaze quoted in Michael Renner, "Forging Environmental Alliances," *World Watch*, November/December 1989.
4. U.S. Department of State, Office of the Spokesman, "The Secretary of State: Memorandum to All Under and Assistant Secretaries," Washington, D.C., February 14, 1996.
5. Laura Parsons and Ross Hammond, "Structural Adjustment in Bolivia: The World Bank, the IMF and Illegal Drug Production," in the APC electronic conference igc:econ.saps on March 7, 1995; share of Bolivian GNP accounted for by drugs from Ed Ayres, "The Expanding Shadow Economy," *World Watch*, July/August 1996.
6. Value of global drug trade from U.N. Research Institute for Social Development, *States of Disarray: The Social Effects of Globalization* (Geneva: 1995).
7. B. Drummond Ayres Jr., "Flow of Illegal Aliens Rises as the Peso Falls," *New York Times*, February 4, 1995; Anthony DePalma, "Economy Reeling, Mexicans Prepare Tough New Steps," *New York Times*, February 26, 1995; David E. Sanger, "Peso Rescue Sets New Limits on Mexico," *New York Times*, February 22, 1995.
8. David E. Sanger, "Chinese Sought in Plot to Import Arms to the U.S.," *New York Times*, May 23, 1996; David E. Sanger, "Chinese Arms Seized in Undercover Inquiry," *New York Times*, May 24, 1996.
9. Lothar Brock, "Abschied von einer großen Idee?," *Der Überblick*, September 1995.
10. Kaplan, op. cit. note 1.
11. Steven Greenhouse, "The Greening of U.S. Diplomacy: Focus on Ecology," *New York Times*, October 9, 1995.
12. Ibid.
13. U.S. Department of State, Office of the Spokesman, "Address by Secretary of State Warren Christopher: American Diplomacy and the Global Environmental Challenges of the 21st Century," Stanford University, Palo Alto, Calif., April 9, 1996; U.S. Department of State, op. cit. note 4.
14. Commission on North-South Issues, *North-South: A*

Programme for Survival (London: Pan Books, Ltd., 1980); Independent Commission on Disarmament and Security Issues, *Common Security* (London: Pan Books, Ltd., 1982); World Commission on Environment and Development, *Our Common Future* (New York: Oxford University Press, 1987); Commission on Global Governance, *Our Global Neighborhood* (New York: Oxford University Press, 1995).

Index

Afghanistan, 159
agriculture
cropland loss, 36–43, 48–50, 68–72
drought, 50–51, 106
economic inequality, 88
import tariffs, 89
land distribution, 83–85, 124–27, 146
meat production, 68–69
air pollution, 141
aquifer depletion, 41–42
Aral Sea basin, 35–36
Arias, Oscar, 164–65, 180
arms
buy-back programs, 162–63
control treaties, 160–63
funding, 176–78, 180
global proliferation, 17–18, 26, 156–60, 193–94
international registry, 163–66
national militarization, 29–30
nuclear weapons, 194
see also war

Bangladesh
flood damage, 40,49, 63
global warming effects, 49
low-income micro-loans, 147–48
migration violence, 97–98
water resource conflicts, 62–64
Barry, Tom, 124–25, 128, 130
Bello, Walden, 86
Böge, Volker, 56
Bolivia, 76, 192
Bosnia, 159–60, 171
Bougainville conflict, 52–56, 59, 141
Brandt Commission, 197

Brazil
grain imports, 89
land redistribution policy, 146–47
social inequity, 131
Bread for the World (U.S.), 86
Bretton Woods institutions, 143–45
Brundtland, Gro Harlem, 186
Brundtland Commission, 197

Canada, 43, 72–73
cancer, 47
Cárdenas, Lázaro, 126
Cardoso, Fernando Henrique, 147
Carnegie Commission on Preventing Deadly Conflict, 170
Carter, Jimmy, 169
Carter Center, Conflict Resolution Program, 169
Central Intelligence Agency (U.S.), 19, 159
Chiapas conflict, 122–32
Childers, Erskine, 167
children, 21, 94
Chile, 77
China
cropland loss, 39
industrialization, 140
land redistribution policy, 146
migration patterns, 98, 107–08
social inequity, 132
water resource conflicts, 64–66
chlorofluorocarbons, 47, 138, 181–82
Chr. Michelsen Institute (Norway), 69
Christopher, Warren, 192, 196
civil society organizations, 137, 151–53
climate
Framework Convention on Climate Change, 138
Intergovernmental Panel on Climate Change (U.N.), 48, 109
resource costs, 178
see also global warming
Clinton administration, 158, 195
coca production, 192
Codevilla, Angelo, 195
coffee production, 120, 124, 128, 192
cold war, 17–19, 22
Collier, George, 123, 126–27, 129
Colombia
arms proliferation, 159
economic decline, 88
illegal drug crops, 192
Commission on Global Governance, 152–53, 168, 187, 197
Commission on Sustainable Development (U.N.), 139
Committee to Protect Journalists, 170
Conference on Security and Co-operation in Europe, 165, 168
Conference on Security, Stability, Development, and Co-operation in Africa, 169
Conflict Prevention Center (Austria), 168
conflicts, *see specific conflicts;* war
Costa Rica, 87
Council for a Livable World Education Fund, 19
Council on Foreign Relations (U.S.), 174
Cramer, Deborah, 45
cropland loss
causes, 38
desertification, 36–37, 39–41
economic effects, 36–39
erosion, 37–39
rising seas, 48–50

salinization, 42–43
shifted cultivators, 37, 39

dams
Farakka Barrage (India), 63–64
Grand Anatolia Project (Turkey), 62
Sardar-Sarovar Narmada Dam Project (India), 66–68
Syrian Dam Project, 62
Three Gorges Dam Project (China), 65–66
water resource conflicts, 60–68
Dean, Jonathan, 172
Defense Intelligence Agency (U.S.), 158
deforestation, 38–40, 106, 110, 125
demilitarization, *see* disarmament
desertification, 36–37, 39–41
Desertification Trust Fund, 181
developing countries
arms control, 160–63
debt relief, 184–85
economic inequality, 78–89
foreign debt, 86
industrialization policy, 142
social development, 142–45
social instability, 95–96
sustainable development funding, 180–82
see also specific countries
disarmament, 30–31, 160–63, 180, 197
drought, 50–51, 106
drugs, 29, 192

Earth Summit, 139, 178
Echeverria, Luis, 127
Economic Commission for Latin America and the Caribbean (U.N.), 77
economy
deregulation effects, 149–50
global inequality, 25, 76–96
globalization, 27–29
human security budget, 173–88
import tariffs, 89
job inequities, 76, 78, 89–94
land degradation effects, 36–39
low-income micro-loans, 147–48
Mexican peso crisis, 193
migrant remittances, 99–100
unemployment rates, 91, 94
world market integration, 143–45
Egypt, 49, 51, 62
employment, *see* jobs
environment
conflict over, 52–75
deforestation, 38–40, 106, 110, 125
degradation effects, 53–54, 57–58, 68–72
environmental refugees, 105–09
environmental stress, 35–51
global warming, 47–51, 108–09, 139–40
international policy, 53–54, 138–42, 196
military impact, 155
mining, 52, 55–56, 141
program funding, 175, 178–80
resource depletion, 25–26, 140–41
Toxics Release Inventory, 151
Environmental Protection Agency (U.S.), 196
erosion, 37–39
Ethiopia, 62, 106
European Council of Ministers, 165
European Union, 73
exclusive economic zones, 72

farmland, *see* cropland loss
fisheries
Aral Sea basin, 35–36

decline, 43–46
environmental conflicts, 72–74
exclusive economic zones, 72
job losses, 44–45, 73
radiation effects, 47
floods
deforestation effects, 39–40
rising water levels, 48–50
water resource conflicts, 63
food
Brazilian grain imports, 89
drought effects, 50–51
fish consumption, 43, 45
humanitarian aid, 174–75
scarcity, 45–46
Food and Agriculture Organization (U.N.), 43
Fordham Institute for Innovation in Social Policy, 83
forests, 38–40, 106, 110, 125, 151
Forum on Environment and Development (Germany), 139
Framework Convention on Climate Change, 138
France, 91
Friedrich Ebert Foundation (Germany), 170

Gabon, 98
Germany
arms sales, 157
migration violence, 98
unemployment rate, 91
Gleick, Peter, 60, 65
Global Environment Facility, 182
globalization
deregulation effects, 149–50
economic inequality, 78–83
human security partnership, 189–98
policy needs, 137–42
trends, 27–29
global warming
effects, 48–51
emission policy, 139–40
environmental refugees, 108–09
projected rise, 47–48
Goldstone, Jack, 108
Gonzalez, Javier, 76
Goose, Stephen, 26
grain imports, 89
Grameen Bank, 147–48
greenhouse effect, *see* global warming
Greenpeace, 74

Habyarimana, Juvénal, 114, 116, 119–21
Hafiz, M. Abdul, 40, 62–63
Hartung, William, 158–59
Hazarika, Sanjoy, 63
Henry Stimson Center (U.S.), 176
Holt, Victoria, 19
Homer-Dixon, Thomas, 117, 128
Honduras, 106–07
Howard, Philip, 128
Human Development Report
global conflict prevention, 167
human security, 30
rich-poor disparities, 80, 82, 88, 183–84
sustainable development, 18
humanitarian aid, 21, 174–75
Human Rights Watch (U.S.), 26, 121
human security
cold war impact, 18
economic security, 145–50
environmental policy, 138–42
financing, 173–88
global partnership, 189–98
information exchange, 150–53, 167
international peace enhancement, 154–72
nongovernmental organizations'

role, 151–53
policy, 135–53
transformation of, 17–31
Hungary, 93
Huntington, Samuel, 23
Hussein, Saddam, 171

import tariffs, 89
Independent Commission on Disarmament and Security Issues, 30
India
economic decline, 88
flood damage, 39–40,51
global warming effects, 51
migration violence, 97–98
Sangam organic farming project, 151
social inequity, 131
water resource conflicts, 62–64, 66–68
industrialization, 140, 142
inequality, 25, 76–96
Institute for Democracy and Electoral Assistance (Sweden), 170
Institute for Development Studies (U.K.), 83–84
Institute for Food and Development Policy (U.S.), 86
Institute for Resource and Security Studies (U.S.), 170
Institute of Peace (U.S), 23–24
Intergovernmental Panel on Climate Change (U.N.), 48, 109
Interhemispheric Resource Center (U.S.), 124–25, 128
International Alert, 170
International Disarmament Fund, 180
International Fund for Social Development (proposed), 184
International Labour Organisation, 93–94, 104
International Monetary Fund
economic adjustment programs, 85–86
foreign debt, 144–45
Sudanese famine, 70–71
world market integration, 143–45
International Peace Research Institute (Norway), 21–22
International Rivers Network (U.S.), 65–66
Iraq, 163
Islam, Nahid, 40, 62–63
Israel, 60–62
Italy, 91

Jacobson, Jodi, 108
Japan, 74, 91
jobs
economic inequality, 76, 78
fisheries decline, 44–45, 73
global inequity, 89–94

Kane, Hal, 105
Kaplan, Robert, 189, 195
Kazakstan, 35
Kuwait, 163, 171

land
cropland loss, 36–43, 48–50, 68–72
degradation, 36–39, 68–72
distribution inequalities, 83–85, 124–27, 146
Latin America
economic decline, 87
economic inequality, 76–83
land distribution, 83–84
land reform, 146–47
see also specific countries
Law of the Sea, 72
Lebanon, 159
Liberia, 103
Libya, 98

Co-operation in Europe, 168–69
Oxfam International, 145
ozone depletion
funding programs, 181–82
global effects, 46–47
international policy, 138

Pacific Institute for Studies in Development, Environment and Security (U.S.), 60
Pakistan, 159
Palestine, 62
Palme, Olof, 30
Palme Commission, 197
Papua New Guinea, 52–56, 59, 141
Partido Revolucionario Institucional (Mexico), 126, 128–29
peace, 154–72
Percival, Valerie, 117
Peru, 151, 192
Poland, 93
policy
environmental stress management, 53–54, 138–42, 196
globalization, 137–42
human security, 138–53
industrialization, 142
land redistribution, 146–47
migrants, 98–99
ozone depletion, 138
social policy, 142–45
pollution
air pollution controls, 141
fisheries decline, 44
Toxics Release Inventory, 151
post–cold war era, 17–19, 22
Project on Defense Alternatives (U.S.), 176–78
Project on Environment, Population, and Security (Canada), 117, 128

Red Cross, 174
Reding, Andrew, 130
refugees, *see* migration
Register of Conventional Arms (U.N.), 163–66
Responsible Care program, 149–50
Rosenau, James, 22, 27
Russia
arms sales, 157
fishery conflicts, 74
migration violence, 98
ozone depletion funding, 182
social inequity, 132
unemployment rate, 93
Rwanda, 26
National Republican Movement for Democracy and Development, 121
refugees, 102, 110
Tutsi/Hutu conflict, 114–22, 131

Sachs, Aaron, 140–41
Salinas de Gortari, Carlos, 123, 127–28
Salvador, 106–07
Sangam organic farming project (India), 151
Sanger, David, 193
Saro-Wiwa, Ken, 58
sea level rise, 108–09
security, *see* human security
Senegal, 88
separatists, 24–25
Shell Oil Company, 57–58
Shevardnaze, Eduard, 190
shifted cultivators, 37, 39
Sierra Leone, 103
Sivard, Ruth, 155
skin cancer, 47
Smil, Vaclav, 64–65, 107–08
Smith, Dan, 21–22
Smyth, Frank, 26

159
Defense Intelligence Agency, 158
drought effects, 50
Environmental Protection Agency, 196
illegal drug trade, 192
illegal immigrants, 87, 98, 193
Institute of Peace, 23–24
job displacement, 90–91
Mexican bailout, 193
National Endowment for Democracy, 170
North American Free Trade Agreement, 122, 127
right-to-know legislation, 151
social inequity, 132

Vietnam, 159

war
casualties, 20–21
globalization effects, 28
internal conflicts, 19–21
international peace enhancement, 154–72
national conflicts, 20, 23–24
resource conflicts, 52–75, 141
see also arms; *specific conflicts*
Warsaw Pact, 166
water
floods, 48–50, 63
resource competition, 59–68
rising sea levels, 48–50
scarcity, 41–42
wealth, 79–83, 136
weapons, *see* arms
Weiner, Myron, 111
Winston Foundation for World Peace (U.S.), 112
Wirth, Timothy, 195
Woods Hole Oceanographic Institution (U.S.), 108
Woolsey, James, 19
World Bank
dam effects, 66
economic adjustment programs, 85–86
foreign debt, 144–45
Global Environment Facility funding, 182
indebted nations, 185
Mexican anti-poverty program, 128
micro-loans, 148
Rwanda conflict, 118, 120
Sudanese mechanized farming, 69–71
world market integration, 143–45
World Conservation Union, 71–72
World Development Movement (U.K.), 183
World Food Programme (U.N.), 174
World Policy Institute (U.S.), 130, 158–59
World Social Summit, 135, 142–43, 153, 183–84

Yanesha Sustainable Forestry Cooperative (Peru), 151
Yunus, Mohammad, 147–48

Zaire, 110, 132
Zapatista National Liberation Army (Mexico), 122–32
Zedillo, Ernesto, 130
Zimbabwe, 160

ABOUT THE AUTHOR

Michael Renner is a Senior Researcher at the Worldwatch Institute. His research and writing deals with disarmament, arms conversion, peacekeeping, the military-environment relationship, and new conceptions of security. During his nine years with the Institute, he has written six Worldwatch Papers including, most recently, *Budgeting for Disarmament: The Costs of War and Peace*. He has also been a contributor to the other Worldwatch publications, *State of the World*, *Vital Signs*, and *World Watch* magazine. Mr. Renner is the author of *Economic Adjustments after the Cold War: Strategies for Conversion*, a study commissioned by the United Nations Institute for Disarmament Research (UNIDIR) in Geneva, Switzerland, and published in 1992 by Dartmouth Publishing Co. in Aldershot, Britain. Before joining Worldwatch in 1987, he was a Corliss Lamont Fellow in Economic Conversion at Columbia University from 1986 to 1987 and a Research Associate at the World Policy Institute in New York City from 1984 to 1986. Renner is a native of the Federal Republic of Germany. He holds degrees in international relations and political science from the Universities of Amsterdam, the Netherlands, and Konstanz, Germany.